AF530253

Solving Problems in Thermodynamics

(GATE XE)

The Authors

Dr. Kshirod Kumar dash, Ph.D., is working as an Associate Professor in the Department of Food Processing Technology, Ghani Khan Choudhury Institute of Engineering & Technology (GKCIET), Malda, West Bengal, India. Previously, he worked as an Assistant Professor in the Department of Food Engineering and Technology, Tezpur University, Tezpur, Assam. Dr. Dash completed his B.Tech in Agricultural Engineering from College of Agricultural Engineering & Technology, OUAT, Bhubaneswar; M.Tech. in Dairy and Food Engineering from IIT Kharagpur; and Ph.D. in Food Process Engineering from IIT Kharagpur. Dr. Dash has obtained his Post-Doc from Ohio State University, Columbus, Ohio, USA. He has guided several postgraduate and doctoral students and published more than 50 publications in peer-reviewed international journals and book chapters in edited volumes. He is a life member of different professional bodies in India and a member of IFT, USA. The research interests of Dr. Dash include Food Process Engineering, Unit Operations in Food Processing, Fluid Mechanics, Thermodynamics, Heat and mass transfer in food processing.

Dr. Paragmoni Kalita is presently working as an Assistant Professor in the Department of Mechanical Engineering, Tezpur University. Prior to joining his service at Tezpur University, Dr. Kalita also had worked in National Engineering Industries Limited, Defence Research and Development Organization (DRDO), Numarligarh Refinery Limited (NRL), and Oil India Limited (OIL). Dr. Kalita completed his B.E. in Mechanical Engineering from erstwhile MNREC (presently MNNIT), Allahabad in the year 2001; M.Tech. in Mechanical Engineering from erstwhile BHU (presently IIT BHU), Varanasi in the year 2004; and Ph.D. in Mechanical Engineering from IIT Guwahati in the year 2017. He is the recipient of the BHU Medal for standing first in the M.Tech (Mechanical Engineering) Examination for 2004. His areas of expertise are Computational Fluid Dynamics and Numerical Heat Transfer. Dr. Kalita has presented his research papers in various reputed national and international conferences. His research works also got published in prestigious international journals, including the International Journal of Applied and Computational Mathematics,Thermodynamics, Computers & Fluids, Journal of Computational Physics, and International Journal of Numerical Methods in Heat & Fluid Flow.

Solving Problems in Thermodynamics

(GATE XE)

Kshirod Kumar Dash
Paragmoni Kalita

2022
Astral Cracker
A Division of
Astral International Pvt. Ltd.
New Delhi – 110 002

© 2022 AUTHORS

ISBN: 978-93-5461-061-5 (PB)

Disclaimer:

Every possible effort has been made to ensure that the information contained in this book is accurate at the time of going to press, and the publisher and author cannot accept responsibility for any errors or omissions, however caused. No responsibility for loss or damage occasioned to any person acting, or refraining from action, as a result of the material in this publication can be accepted by the editor, the publisher or the author. The Publisher is not associated with any product or vendor mentioned in the book. The contents of this work are intended to further general scientific research, understanding and discussion only. Readers should consult with a specialist where appropriate.

Every effort has been made to trace the owners of copyright material used in this book, if any. The author and the publisher will be grateful for any omission brought to their notice for acknowledgment in the future editions of the book.

All Rights reserved under International Copyright Conventions. No part of this publication may be reproduced, stored in a retrieval system, or transmitted in any form or by any means, electronic, mechanical, photocopying, recording or otherwise without the prior written consent of the publisher and the copyright owner.

Published by : **Astral Cracker**
A Division of
Astral International Pvt. Ltd.
– ISO 9001 : 2015 Certified Company –
4736/23, Ansari Road, Darya Ganj,
New Delhi-110 002
Ph. 011-43549197, 23278134
E-mail: info@astralint.com
Website: www.astralint.com

Preface

The objective of Solving Problems in Thermodynamics (GATE XE) is to guide potential candidates preparing for GATE in solving complex problems on thermodynamics through the development of clear concepts on the subject matter. The book provides the solutions to the Thermodynamics (XE-E) sections of GATE Engineering Sciences (XE) question papers from 2009 to 2021. Accordingly, efforts have been put in not just to present the solutions or final answers to the questions, but also on elaborating the underlying concepts and detailed methodology wherever necessary. The book covers problems in basic concepts of thermodynamics,Zeroth law of thermodynamics, first and second laws of thermodynamics, thermodynamic relations, properties of pure substances, properties of ideal gases and ideal gas mixtures, and thermodynamic cycles. This book can be useful for candidates preparing for GATE Engineering Sciences (XE), GATE Food Technology, GATE Mechanical Engineering, and GATE Agricultural Engineering. The book will also be beneficial to industry and academic experts from mechanical engineering, food engineering, and related fields.

We express our thanks to all the students and faculty of the Department of Food Processing Technology, Ghani Khan Choudhury Institute of Engineering & Technology (GKCIET), Malda, West Bengal; Department of Food Engineering and Technology, Tezpur University; and Department of Mechanical Engineering, Tezpur University for their continuous encouragement to publish this book. We do hope that this book will be a valuable resource for candidates preparing for the GATE examinations and the knowledge presented in the book will strengthen the concepts of the readers on thermodynamics.

Dr. Kshirod Kumar Dash
Dr. Paragmoni Kalita

CONTENTS

GATE 2021
THERMODYNAMICS

Q.1 A refrigerator working on a reversed Carnot cycle has a Coefficient of Performance (COP) of 4. If it works as a heat pump and consumes work in put of 1 kW, the heating effect will be:

(A) 1 kW (B) 4 kW
(C) 5 kW (D) 6 kW

Solution

Given:

COP for cooling, $COP_{heating} = 4$

The work input when acting as a heat pump, $W_{in} = 1kW$

Now, the COP for heating is given by

$COP_{heating} = 1 + COP_{heating} = 1 + 4 = 5$

Also, $COP_{heating} = \frac{Q_{heating}}{W_{in}}$, where $Q_{heating}$ is the heating effect created by the heat pump.

$\therefore Q_{heating} = COP_{heating} \times W_{in} = 5 \times 1 \text{ kW} = 5 \text{ kW}$

Hence, the answer is (C).

Q.2 The liquid phase of pure substance is termed as__________________, if its temperature is lower than the saturation temperature corresponding to its pressure.

(A) super-heatedliquid (B) sub-cooledliquid
(C) metastableliquid (D) flashingliquid

Solution to Q.2

The liquid phase of a substance is termed as **super-cooled liquid**, if its temperature is lower than the saturation temperature corresponding to its pressure *P*.

Hence, the answer is (B).

Q.3 Two air streams of mass flow rates m_1 and m_2 enter a mixing chamber and exit after perfect mixing. The corresponding temperatures of the inlet streams are T_1 and T_2, respectively. Heat loss rate from the mixing chamber to the surrounding is Q. Assume that the process is steady, specific heat capacity is constant, and air behaves as an ideal gas. Identify the correct expression for the final exit temperature T_3 after mixing. The mass specific heat capacities of the gas at constant volume and constant pressure are c_v and c_p, respectively.

Neglect the bulk kinetic and potential energies of the streams

(A) $T_3 = \dfrac{m_1 T_1 + m_2 T_2}{m_1 + m_2} - \dfrac{Q}{C_v(m_1 + m_2)}$ (B) $T_3 = \dfrac{m_1 T_1 + m_2 T_2}{m_1 + m_2} + \dfrac{Q}{C_p(m_1 + m_2)}$

(C) $T_3 = \dfrac{m_1 T_1 + m_2 T_2}{m_1 + m_2} - \dfrac{Q}{C_p(m_1 + m_2)}$ (D) $T_3 = \dfrac{m_1 T_1 + m_2 T_2}{m_1 + m_2} + \dfrac{Q}{C_v(m_1 + m_2)}$

Solution

The steady flow energy equation for the mixing process is given by

$$\dot{m}_1 h_1 + \dot{m}_2 h_2 - \dot{Q} = (\dot{m}_1 + \dot{m}_2) h_3, \tag{i}$$

where h_1 and h_2 are the specific enthalpies of the two mixing air streams with mass flow rates $\dot{m}_1$ and $\dot{m}_2$, respectively, and h_3 is the specific enthalpy of the air mixture at the exit. Now, considering air as a perfect gas, Eq. (i) becomes

$$\dot{m}_1 C_p T_1 + \dot{m}_2 C_p T_2 - \dot{Q} = (\dot{m}_1 + \dot{m}_2) C_p T_3$$

$$\therefore \quad T_3 = \frac{\dot{m}_1 T_1 + \dot{m}_2 T_2}{\dot{m}_1 + \dot{m}_2} - \frac{\dot{Q}}{C_p(\dot{m}_1 + \dot{m}_2)}$$

Hence, the answer is (C).

Q.4 If

h is the mass specific enthalpy,

s is the mass specific entropy,

P is the pressure,

T is the temperature,

C_V is the mass specific heat at constant volume,

C_P is the mass specific heat at constant pressure,

β is the coefficient of thermal expansion,

v is the mass specific volume,

k is the isothermal compressibility,

then the partial derivative $\left(\dfrac{\partial h}{\partial s}\right)_P =$

(A) $\left(T - \frac{1}{\beta}\right)\left(\frac{C_P}{C_V}\right)$ (B) $\left(T - \frac{1}{\beta}\right)$

(C) $T\left(1 - \frac{v\beta}{kCv}\right)_r$ (D) T

Solution

The first law and second law of thermodynamics together give the thermodynamic identity:

$$Tds = dh - vdP \quad \text{(i)}$$

From Eq. (i) we get

$$\left(\frac{\partial h}{\partial s}\right)_p = T$$

Hence, the answer is (D).

Q5. If
***v* is the mass specific volume,**
***s* is the mass specific entropy,**
***P* is the pressure,**
***T* is the temperature,**
then using Maxwell relations, $\left(\frac{\partial S}{\partial P}\right)_T =$

(A) $\left(\frac{\partial v}{\partial T}\right)_P$ (B) $-\left(\frac{\partial v}{\partial T}\right)_P$

(C) $\left(\frac{\partial v}{\partial T}\right)_S$ (D) $-\left(\frac{\partial v}{\partial T}\right)_S$

Solution

The differential Gibbs function is given by

$$dg = vdP - sdT$$

Since the Gibbs function is an exact differential, therefore we can write from Eq. (i) that

$$\left(\frac{\partial s}{\partial P}\right)_T = -\left(\frac{\partial v}{\partial T}\right)_p$$

Hence, the answer is (B).

Q.6 A closed system consists of a solution of liquid water and ethanol in equilibrium with its vapours Using the gibbs phase rule, the degree of freedom of the system is:

(A) 0 (B) 1
(C) 2 (D) 3

Solution to Q.6

The Gibbs phase rule relating the number of degrees of freedom F, number of components C, and the number of phased P is

$$F = C - P + 2.$$

For the present problem, we have $C = 2$ and also $P = 2$. Therefore, the Gibbs phase rule gives $F = 2$.

Hence, the answer is (C).

Q.7 For a real gas passing through an insulated throttling valve, the out let temperature of the gas________with respect to the in let temperature.

(A) is always higher (B) is always lower

(C) maybe higher, lower or same (D) is always same

Multiple Select Question (MSQ), Carry ON Emark each (no negative marks).

Solution

The Joule-Thomson coefficient for a substance is given by

$$\mu_J = \left(\frac{\partial T}{\partial P}\right)_h$$

It may be noted that a substance will produce cooling effect due to throttling if its $\mu_J > 0$ and vice-versa. For a real gas the value of μ_J can either be greater than or less than effect. Therefore, the throttling of a real gas may either produce a cooling effect or heating effect depending on whether the value of its Joule-Thomson coefficient is greater than or less than 0.

Hence, the answer is (C).

Q.8 Atmospheric air with Dry Bulb Temperature (DBT) of 24 °C and Relative Humidity of 35%, entering in a circular duct (assume no pressure drop in the duct) is heated by an electrical resistance arrangement inside the duct. The DBT of air measured at the outlet of the duct is equal to 30°C. Considering the flow to be-steady, which of the following statement (s) is (are) correct as-regards to the out let air, with respect to the in let air?

(A) There is no change in the Relative Humidity

(B) There is no change in the Dew Point Temperature

(C) There is no change in the Specific Humidity

(D) There is no change in the Specific Enthalpy

Solution

The given psychrometric process refers to a sensible heating process where the specific humidity ω remains constant, but the relative humidity decreases. However, when specific humidity remains constant, the partial pressure of water vapour in the air-vapour mixture also remains constant. Therefore, the dew point temperature, which is the saturation temperature of water vapour corresponding to its partial pressure in the air-vapour mixture also does not change due to sensible heating.

Hence two options are correct and those are (B), (C).

Q. 9. Numerical Answer Type (NAT), carry ONE mark each (none gative marks).

A cylinder of volume 1 m3 contains a mixture of CO_2 (20% by mol) and O_2 (80% by mol) at 100 kPa and 300 K. This cylinder is connected to a 1 MPa pressure line carrying N_2 at 300 K. The cylinder is filled isothermally till the pressure of gas mixture inside it becomes 500 kPa, and then the filling is stopped. The amount of N_2 gas that has entered the cylinder is ______________ (in mole, 2 decimal places). The universal gas constant is 8.3145 J/(mol K).

Solution to Q.9

Given:

Initial pressure, $P_1 = 100$ kPa $= 100 \times 10^3$ Pa

Final pressure, , $P_2 = 500$ kPa $= 500 \times 10^3$ Pa

Initial mole fraction of CO_2, $(X_1)_{CO_2} = 0.2$

Initial mole fraction of O_2, $(X_1)_{O_2} = 0.8$

Volume of the cylinder, $V = 1$ m^3

The constant temperature maintained, $T = 300$ K

Let the total number of moles of CO_2 and O_2 be n_1. Then from the ideal gas equation of state we have

$$n_1 = \frac{P_1 V}{R_o T} = \frac{100 \times 10^3 \times 1}{8.3145 \times 300} = 40.09 \qquad \text{(i)}$$

Let the total number of moles CO_2, O_2 and N_2 after filling of N_2 be n_2. Then from the ideal gas equation of state, we get

$$n_2 = \frac{P_2 V}{R_0 T} = \frac{500 \times 10^3 \times 1}{8.3145 \times 300} = 200.45 \qquad \text{(ii)}$$

Therefore, the number of moles of N_2 that has entered the cylinder

$= 200.45 - 40.09 = 160.36$

Hence, the answer is 160.36.

Q. 10 – Q. 13 Multiple Choice Question (MCQ), carry TWO marks each (for each wrong answer: – 2/3).

Q. 10. he saturation pressure P_sat of a pure liquid is represented by an equation of the form: ln P_{sat} = A – (B/T), where, A and B are constants, and T is the absolute temperature. For this substance, which of the following expression for specific entropy difference between the saturated vapour and the saturated liquid phase (s_{fg}) is correct?

Note: Subscripts f and g refer to saturated liquid and saturated vapour phases, respectively, and v_{fg} is the specific volume difference between the saturated

vapour and the saturated liquid phases

(A) $S_{fg} = v_{fg}\,\dfrac{BP^{2}_{sat}}{T^2}$ (B) $S_{fg} = v_{fg}\,\dfrac{BP^{2}_{sat}}{T^2}$

(C) $S_{fg} = v_{fg}\,\dfrac{BP_{sat}}{T^3}$ (D) $S_{fg} = v_{fg}\,\dfrac{BP^{3}_{sat}}{T^2}$

Solution

The Clausius-Clapeyron equation gives

$$\frac{dP_{sat}}{dT} = \frac{S_{fg}}{v_{fg}} \quad \text{(i)}$$

Re-arranging Eq. (i) we get

$$s_{fg} = v_{fg}\,\frac{dP_{sat}}{dT} \quad \text{(ii)}$$

The given relation is

$$\ln P_{sat} = A - \frac{B}{T} \quad \text{(iii)}$$

The differentiation of Eq. (iii) yields

$$\frac{dP_{sat}}{dT} = \frac{B}{T^2}\,dT \quad \text{(iv)}$$

$$\frac{dP_{sat}}{dT} = \frac{BP_{sat}}{T^2}\,dT$$

Substituting for $\dfrac{dP_{sat}}{dT}$ from Eq. (iv) in Eq. (ii) we get

$$s_{fg} = v_{fg}\,\frac{BP_{sat}}{T^2} \quad \text{(v)}$$

Hence, the answer is (B).

Q.11 For a refrigeration cycle, the ratio of actual COP to the COP of a reversible refrigerator operating between the same temperature limits is 0.8. The condenser and evaporator temperatures are 51 °C and 30°C, respectively. If the cooling capacity of the plant is 2.4 kW, then the power input to the refrigerator is:

(COP: Coefficient of Performance)

(A) 1.00 kW (B) 1.22 kW

(C) 1.25 kW (D) 2.08 kW

Solution

Given:

$COP_{actual} = 0.8 COP_{reversible}$

Evaporator temperature, $T_0 = 243$ K

Condenser temperature, $T_k = 324$ K

Refrigeration capacity, $Q_0 = 2.4$ kW

Now the COP of the reversible refrigerator,

$$COP_{reversible} = \frac{T_0}{T_k - T_0} = \frac{243}{324 - 243} = 3$$

Therefore, the COP of the actual refrigerator will be

$$COP_{actual} = 0.8COP_{reversible} = 0.8 \times 3 = 2.4$$

Now, if the power input to the actual refrigerator is W_{in}, then the COP of the actual refrigerator is also given by

$$COP_{actual} = \frac{Q_0}{W_{in}}$$

$$\therefore W_{in} = \frac{Q_0}{COP_{actual}} = \frac{2.4}{2.4}\ \text{kW} = 1\ \text{kW}$$

Hence, the answer is (A).

Q.12. Two identical pressure cookers, Cooker A and Cooker B, each having a total internal capacity of 6 litres are available. Cooker A is filled with 2 litres of liquid water at 110 C and Cooker B is filled with 4 litres of liquid water at 110 C. The remaining space in both the cookers is filled with saturated water vapour in equilibrium with the liquid water. If *g* represents the specific Gibbs free energy, and subscripts *v* and *l* represent the saturated vapour and the saturated liquid phases, respectively, which of the following expressions is correct?

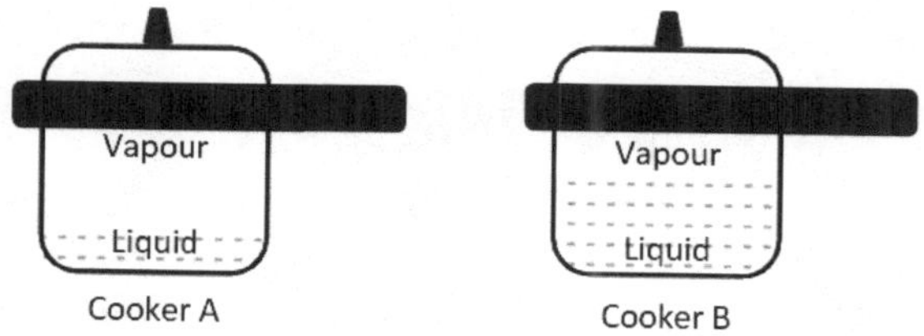

(A) $g_{v,A} > g_{l,B}$ (B) $g_{v,A} < g_{l,B}$

(C) $g_{v,A} = g_{l,B}$ (D) $g_{l,B} = 2g_{l,A}$

Solution

In both the pressure cookers *A* and *B*, the liquid and vapour phases of water exist in equilibrium at the same temperature. Now, for equilibrium between the liquid and vapour phases of a substance, if g_l and g_v are the specific Gibbs free energy for the liquid and vapour phases, respectively, then it can be shown that at a given temperature and hence at a given pressure

$$g_v = g_l$$

Since in the present case, both the pressure cookers are at the same temperature of 110°C, therefore, we can write

$$g_{v,A} = g_{l,B}$$

Hence, the answer is (C).

Q. 13. Four different Entropy (S) - Temperature (T) diagrams, representing liquid to vapour phase transition process of a pure substance in a closed system under constant pressure are shown. The diagram, which correctly represents the process, is:

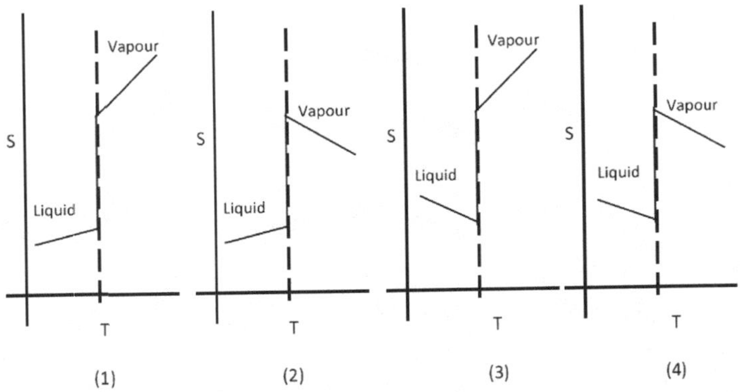

(A) 1 (B) 2
(C) 3 (D) 4

Solution

From thermodynamics relations it can be shown that the partial derivative of specific entropy s w.r.t. the absolute temperature T at a constant pressure P is obtained as

$$\left(\frac{\partial s}{\partial T}\right)_P = \frac{C_P}{T},$$

where C_p is the specific heat at constant pressure. Since both C_p and T are positive numbers, therefore, $\left(\frac{\partial s}{\partial T}\right)_P$ is always greater than 0. Therefore, diagram (1) represents the correct variation of entropy with the temperature at a constant pressure.

Hence, the answer is (A).

Q.14 Air having a mass flow rate of 2 kg/s enters a diffuser at 100 kPa and 30 C, with a velocity of 200 m/s. Exit area of the diffuser is 400 cm2 while the exit temperature of the air is 45 C. The rate of heat loss from the diffuser to the surrounding is 8 kJ/s. The pressure at the diffuser exit is kPa (2 decimal places).

For air, the characteristic gas constant is 287 J/(kgK) and specific heat capacity at constant pressure is 1005 J/(kgK). Assume air to be an ideal gas and the flow in the diffuser is steady.

Solution

Given:

Mass flow rate of air, $\dot{m} = 2$ kg/s

Inlet temperature of air, T_1=(273 + 30) K = 303 K

Inlet velocity of air, $V_1 = 200$ m/s

Exit temperature of air, $T_2 = (273 + 45)$ K = 318 K

Exit area of the diffuser, $A_2 = 400$ cm^2 = 400×10^{-4} m^2

Rate of heat loss to the surroundings, $\dot{Q}$ = 8 kJ/s = 8000 W

Characteristic gas constant, $R = 287$ J/(kgK)

Specific heat at constant pressure, $C_p = 1005$ J/(kgK)

Now, if h_1 and h_2 are the specific enthalpies of air at the inlet and outlet of the diffuser, respectively, then the steady flow energy equation gives

$$\dot{m}\left(h_1 + \frac{1}{2}V_1^2\right) = \dot{m}\left(h_2 + \frac{1}{2}V_2^2\right) + \dot{Q}$$

or,

$$\dot{m}\left(C_p T_1 + \frac{1}{2}V_1^2\right) = \dot{m}\left(C_p T_2 + \frac{1}{2}V_2^2\right) + \dot{Q}$$

or,

$$V_2 = \sqrt{2\left\{C_p (T_1 - T_2) + \frac{1}{2}V_1^2 - \frac{\dot{Q}}{\dot{m}}\right\}}$$

or,

$$V_2 = \sqrt{2\left\{1005 \times (303 - 318) + \frac{1}{2} \times 200^2 - \frac{8000}{2}\right\}} \text{ m/s} = 43.01 \text{ m/s}$$

Now, if the density at outlet is ρ_2, then

$\rho_2 A_2 V_2 = \dot{m}$

or, $$\rho_2 = \frac{\dot{m}}{A_2 V_2}$$

$$\therefore \quad \rho_2 = \frac{2}{400 - 10^{-4} \times 43.01} \text{ kg/m}^3 = 1.163 \text{ kg/m}^3$$

If the pressure at the outlet of the diffuser is P_2, then the ideal gas equation of state gives

$P_2 = \rho_2 R T_2 = 1.163 \times 287 \times 318$ Pa = 106.14 kPa

Hence, the answer is 106.14.

Q 15. For the Refrigerant *R*-134 (at 1 MPa and 50°C), the difference between the specific volume computed by assuming it to be an ideal gas and its actual specific volume is: $v_{ideal} - v_{actual}$ = 4.539 100^{-3} m^3/kg. If the compressibility factor associated with this state is Z = 0.84, then $v_{com} - v_{actual}$ = __________ 10^{-3} m^3/kg (3 decimal places).

Here v_{com} is the specific volume calculated using the compressibility factor. For Refrigerant *R*-134 (at 1 MPa and 50 C):

The characteristic gas constant: 0.0815 kJ/(kgK), The critical pressure and temperature are, respectively, P_{cr} = 4.059 MPa and T_{cr} = 374.2 K.

Solution

Given:

Pressure, $P = 10^6$ Pa

Temperature, $T = (273 + 50)$ K = 323 K

Compressibility factor, $Z = 0.84$

Characteristic gas constant, $R = 0.0815$ kJ/(kg.K) = 81.5 J/(kg.K)

$V_{ideal} - V_{actual} = 4.529 \times 10^{-3}$ m³/kg

Now,

$$V_{ideal} = \frac{RT}{P} = \frac{81.5 \times 323}{10^6} \text{ m}^3/\text{kg} = 0.02633 \text{ m}^3/\text{kg}$$

Therefore,

$$V_{actual} = (V_{ideal} - 4.529 \times 10\text{–}3) \text{ m3/kg}$$

$$= (0.02633 - 4.529 \times 10^{-3}) \text{ m}^3/\text{kg} = 0.0218 \text{ m}^3/\text{kg}$$

The specific volume using the compressibility factor is obtained as

$$v_{com} = Zv_{ideal} = 0.84 \times 0.02633 \text{ m}^3/\text{kg} = 0.0221 \text{ m}^3/\text{kg}$$

Accordingly, we can now calculate the value of $v_{com} - v_{actual}$ as

$$v_{com} - v_{actual} = (0.0221 - 0.0218)\text{m}^3/\text{kg} = 0.3 \times 10^{-3} \text{ m}^3/\text{kg}$$

Hence, the answer is 0.3.

Q.16. A mixture of air and water vapour enters a steady-flow adiabatic saturator at 50°C and 100 kPa. It leaves the saturator in a completely saturated state at temperature of 25 C and pressure of 100 kPa. Liquid water enters the saturator at 25°C. If air is considered to be an ideal gas, with constant specific heat capacity, the relative humidity of the air entering the saturator is _______________ % (1 decimal place).

Use the following data:

at 25 C h_f = 104.87 kJ/kg, h_g = 2547.17 kJ/kg, P_{sat} = 3.161 kPa

at 50 C h_f = 209.31 kJ/kg, h_g = 2592.06 kJ/kg, P_{sat} = 12.335 kPa

Specific heat capacity of air at constant pressure C_p = 1.005 kJ/(kgK)

Solution

Given:

DBT of inlet air, $t = 50$°C

Adiabatic saturation temperature, $t^* = 25$°C

The atmospheric pressure, $P = 100$ kPa

Specific heat of air at constant pressure, Cp = 1.005 kJ/(kgK)

$h_f|_{t^*} = 104.87$ kJ/kg

$h_g|_{t^*} = 2547.17$ kJ/kg

$P_{sat}|_{t^*} = 3.161$ kPa

$h_f|_t = 209.31$ kJ/kg

$h_g|_t = 2592.06$ kJ/kg

$P_{sat}|_t = 12.335$ kPa

Now the specific humidity of the adiabatically saturated air is given by

$$\omega^* = 0.622 \frac{P_{sat}|_{t^*}}{P - P_{sat}|_{t^*}} = 0.622 \times \frac{3.161}{100 - 3.161} = 0.0203$$

The specific humidity of the inlet air is expressed in terms of the adiabatic saturation temperature and other quantities as

$$\omega = \frac{1.005 \times (25 - 50) + 0.0203 \times (2547.17 - 104.87)}{2592.06 - 104.87} = 0.0098$$

If the partial pressure of water vapour in the inlet air is $P_v|_t$ then

$$\omega = 0.622 \frac{P_v|_t}{P - P_v|_t}$$

$$\therefore P_v|_t = \frac{\omega P}{\omega + 0.622} = \frac{0.0098 \times 100}{0.0098 + 0.622} = \text{kPa} = 1.1511\text{kPa}$$

Therefore, the relative humidity of the air entering the adiabatic saturator

$$\varphi = \frac{P_v|_t}{P_{sat}|_t} \times 100\% = 12.58\%$$

Hence, the answer is 12.58

Q. 17. Air at a pressure of 1 MPa and 300 K is flowing in a pipe. An insulated evacuated rigid tank is connected to this pipe through an insulated valve. The volume of the tank is 1 m³. The valve is opened and the tank is filled with air until the pressure in the tank is 1 MPa. Subsequently, the valve is closed.

Consider air to be an ideal gas and neglect bulk kinetic and potential energy. The final temperature of air in the tank is __________ K (1 decimal place).

Specific heat capacity of air at constant pressure C_p = 1.005 kJ/(kgK) and characteristic gas constant for air = 0.287 kJ/(kgK)

Solution

Given:

Inlet temperature of air, $T_i = 300$ K

Inlet pressure of air, $P_i = 1$ MPa

Initial tank pressure, $P_1 = 0$

Final tank pressure, $P_2 = P_i = 1$ MPa

The specific heat capacity of air at constant pressure, $C_p = 1.005$ kJ/(kgK)

The characteristic gas constant for air, $R = 0.287$ kJ/(kgK)

Therefore, the specific heat capacity of air at constant volume,

$$C_v = C_p - R = 0.718 \text{ kJ/(kgK)}$$

Also, the ratio of specific heat capacities at constant pressure and constant volume is obtained as

$$\gamma = \frac{C_p}{C_v} = \frac{1.005}{0.718} = 1.4$$

It is also given that the tank is initially evacuated, i.e., the mass of the tank, $m_{CV}|1 = 0$

The work done and heat transfer during the process are zero.

Let the final temperature of air inside the tank be T_2

Therefore, the energy balance equation for the filling process is

$$m_{CV}|_2 u_2 = h_i m_{CV}|_2, \tag{i}$$

where $m_{CV}|_2$ is the final mass of ais inside the tank, u_2 is final the specific internal energy of air inside the tank, and h_i is the specific enthalpy of the inlet air.

Since air is considered as a perfect gas, hence Eq. (ii) get simplified after dividing both sides of the equation with $m_{CV}|_2$:

$$C_v T_2 = C_p T_i$$

$$T_2 = \frac{C_p}{C_v} T_i = \gamma T_i = 1.4 \times 300 \text{ K} = 420 \text{ K}$$

Hence, the answer is 420.

Q. 18. A cylinder of volume 0.1 m³ is filled with 100 mol of propane (C_3H_8) at 2 MPa. If propane is assumed to obey the van der Waals equation of state, then its temperature is ______________ K (1 decimal place).

The van der Waals constants for propane are: a = 939.2 kPa (m³/kmol)² and b = 0.0905 m³/kmol. The universal gas constant is 8.3145 J/(mol K).

Solution

Given:

Volume, $V = 0.1$ m³

No. of moles, $n = 100$ mol = 0.1 kmol

Pressure, $P = 2$ MPa = 2000 kPa

van der Waal's constants $a = 939.2$ kPa(m³/kmol)², $b = 0.0905$ m³/kmol

Universal gas constant, $R = 8.3145$ J/(mol. K) = 8.3145 kJ/(kmol.K)

The molar volume occupied by the gas

$$v = \frac{V}{n} = \frac{0.1}{0.1} \text{ m}^3\text{/kmol} = 1 \text{ m}^3\text{/kmol}$$

Let the temperature of the gas be T. The van der Waal's equation of state gives

$$T = \frac{\left(p + \frac{a}{v^2}\right)(v - b)}{R} = \frac{\left(2000 + \frac{939.2}{1^2}\right)(1 - 0.0905)}{8.3145} \text{ K} = 321.51 \text{ K}$$

Hence, the answer is 321.5.

Q19. A frictionless piston cylinder device contains 1 kg of an ideal gas. The gas is compressed according to $Pv^{1.3}$ = constant (P is pressure and v is mass specific volume), from 100 kPa, 250 K, till it reaches a temperature of 500 K. The heat transfer from the piston cylinder device to its surroundings is __________ kJ (2 decimal places).

The characteristic gas constant is 287 J/(kgK) and the ratio of specific heat capacities is 1.4.

Solution

Given:

The $P - v$ relation for the polytropic process is $Pv^{1.3}$ = constant

Mass of the gas, $m = 1$ kg

Initial pressure, $P_1 = 100$ kPa

Initial temperature, $T_1 = 250$ K

Initial temperature, $T_2 = -500$ K

The characteristic gas constant of the ideal gas, $R = 287$ J/(kgK)

The ratio of specific heat capacities, $\gamma = 1.4$

Now, from the given polytropic process we have the polytropic index, $n = 1.3$

For a polytropic process Pv^n = constant, the work done by the system during a process 1-2 undergone by an ideal gas is given by

$$W_{1-2} = \frac{mR(T_1 - T_2)}{n - 1}$$

Therefore, for the present process we have

$$W_{1-2} = \frac{1 \times 287 \times (250 - 500)}{1.3 - 1} \text{ J} = -239166.7 \text{ J}$$

If C_v is the specific heat capacity of the gas at constant volume, then the change in internal energy during the process

$$\Delta U = mC_v(T_2 - T_1) = \frac{mR}{\gamma - 1}(T_2 - T_1) = \frac{1 \times 287 \times (500 - 250)}{1.4 - 1} \text{ J} = 179375 \text{ J}$$

Let the heat transfer during the process undergone by the closed system be Q_{1-2}. Then the first law of thermodynamics gives

$$Q_{1-2} = \Delta U + W_{1-2} = (179375 - 239166.7) \text{ J} = -59.79 \text{ kJ}$$

This means that the heat transfer from the piston-cylinder device to the surrounding is 59.79 kJ

Hence, the answer is 59.79.

Q.20. A 0.8 m³ insulated rigid tank contains 1.5 kg of an ideal gas at 100 kPa. Electric work is done on the system until the pressure in the tank rises to 135 kPa. The loss in availability (exergy) associated with the process is ________ kJ (2 decimal places).

For the ideal gas, the characteristic gas constant is 188.9 J/(kgK) and the specific heat capacity at constant volume is 680 J/(kgK). The temperature of the dead state is 298 K.

Solution

Given:

Volume of the tank, $V = 0.8\ \text{m}^3$

Mass of the ideal gas in the tank, $m = 1.5$ kg

Initial pressure, $P_1 = 100$ kPa

Final pressure, $P_2 = 135$ kPa

The temperature of the dead state, $T_0 = 298$ K

The characteristic gas constant, $R = 188.9$k J/(kgK) = 0.1889 kJ/(kgK)

The specific het capacity at constant volume, $C_v = 680$ kJ/(kgK) = 0.68 kJ/(kgK)

Let the initial and final temperature of the gas be T_1 and T_2, respectively.

Let the initial and final entropy of the gas be S_1 and S_2, respectively.

Let the initial and final internal energy of the gas be U_1 and U_2, respectively.

Since the volume of the system remains fixed, hence the loss in availability of the system is given by

$$\Delta A = T_0(S_2 - S_1) \quad \text{...(i)}$$

Now, the initial temperature of the gas is obtained as

$$T_1 = \frac{P_1 V}{mR} = \frac{100 \times 0.8}{1.5 \times 1.889}\ \text{K} = 282.34\ \text{K} \quad \text{...(ii)}$$

Since the mass of the gas and its volume remains constant for the process, therefore the final temperature of the gas is obtained as

$$T_2 = \frac{P_2}{P_1} T_1 = \frac{135}{100} \times 282.34\ \text{K} = 381.16\ \text{K} \quad \text{...(iii)}$$

The change is entropy of the system is obtained as

$$S_2 - S_1 = mC_v \ln\frac{T_2}{T_1} + R\ln\frac{V_2}{V_1} = mC_v \ln\frac{T_2}{T_1}$$

$$S_2 - S_1 = 1.5 \times 0.68 \times \ln\left(\frac{381.16}{282.34}\right)\ \text{kJ/K} = 0.306\ \text{kJ/K} \quad \text{...(iv)}$$

Substituting for $(S_2 - S_1)$ from Eq. (iv) in Eq. (i) we get the loss in availability of the system as

$$\Delta A = 298 \times 0.306\ \text{kJ} = 91.19\ \text{kJ}$$

Hence, the answer is 91.19

Q 21. A rigid tank contains 1.0 kg of pure water consisting of liquid and vapour phases in equilibrium at 10 bar. If the liquid and vapour phase each occupies one half of the volume of the tank, then the net enthalpy of the contents of the tank is ________ kJ (1 decimal place).

For saturated liquid and vapour at 10 bar, the thermodynamic data table provides the following values:

$v_f = 1.127 \;\; 10^{-3}\ \text{m}^3/\text{kg},\ v_g = 194.3 \;\; 10^{-3}\ \text{m}^3/\text{kg},$

$h_f = 762.6\ \text{kJ/kg},\ h_g = 2776.2\ \text{kJ/kg}$

Solution

Given:

Total mass of the liquid water-vapour mixture, $m = 1$ kg

Specific volume of saturate liquid, $v_f = 1.127 \times 10^{-3}\ \text{m}^3/\text{kg}$

Specific volume of saturate vapour, $v_g = 194.3 \times 10^{-3}\ \text{m}^3/\text{kg}$

Specific enthalpy of saturate liquid, $h_f = 762.6$ kJ/kg

Specific enthalpy of saturate vapour, $h_g = 2776.2$ kJ/kg

Let the mass of liquid water and vapour present in the mixture be m_f and m_g, respectively, so that

$$m_f + m_g = m$$

or,
$$m_f = m - m_g = 1 - m_g$$

Since both the liquid and vapour phases occupy one half of the volume of the tank, therefore,

Volume occupied by the liquid = Volume occupied by the vapour

or,
$$m_f v_f = m_g v_g$$

or,
$$m_f v_f = m_g v_g$$

or,
$$(1 - m_g) v_f = m_g v_g$$

or,
$$m_g = \frac{v_f}{v_f + v_g} = \frac{1.127\times10^{-3}}{1.127\times10^{-3} + 194.3\times10^{-3}} = 0.0058\ \text{kg}$$

Therefore, the enthalpy of the mixture,

$$H = m_f h_f + m_g h_g = (1 - m_g) h_f + m_g h_g$$

$$= [(1 - 0.0058) \times 762.6 + 0.0058 \times 2776.2]\ \text{kJ}$$

$$= 774.28\ \text{kJ}$$

Hence, the answer is 774.28.

Q.22. An air-standard Diesel cycle with a compression ratio of 16 takes air at 1 bar and 300 K. If the maximum temperature in the cycle is 2100 K, then the thermal efficiency of the cycle is ______________ % (1 decimal place).

The ratio of the specific heat capacities of air is 1.4.

Solution

Given:

Compression ratio, $r_k = 16$

Intake pressure, $P_1 = 1$ bar

Intake temperature, $T_1 = 300$ K

Maximum temperature in the cycle, $T_3 = 2100$ K

The ratio of specific heat capacities, $\gamma = 1.4$

Let the cut-off ratio be r_c and the temperature at the end of compression be T_2.

If the specific volumes at the intake and at the end of compression are v_1 and v_2, respectively, then $\frac{v_1}{v_2} = r_k$.

Therefore, we have for the isentropic compression process

$$T_2 = T_1\left(\frac{v_2}{v_1}\right)^{\gamma-1} = T_1(r_k)^{(\gamma-1)} = 300 \times (16)(1.4-1)$$

K = 909.43 K

If the cut-off ratio is v_c, then

$$\frac{v_3}{v_2} = r_c = \frac{T_3}{T_2} = \frac{2100}{909.43} = 2.3$$

Therefore, the thermal efficiency of the Diesel cycle is given by

$$\eta_{\text{Diesel}} = 1 - \frac{1}{\gamma}\cdot\frac{1}{r_k^{\gamma-1}}\cdot\frac{r_c^{\gamma}-1}{r_c-1}$$

$$= 1 - \frac{1}{1.4}\times\frac{1}{(16)^{(1.4-1)}}\times\frac{2.3^{1.4}-1}{2.3-1}$$

$$\therefore \eta_{\text{Diesel}} = 60$$

Hence, the answer is 60.

GATE 2020
THERMODYNAMICS

XE: Engineering Sciences - E: Thermodynamics

Q. 1 – Q. 9 carry one mark each.

Q.1 If x and y are two independent intensive properties of a thermodynamic system, then which relation among the followings fails to identify z as another thermodynamic property?

(A) $dz = x\,dy + y\,dx$ (B) $dz = x\,dy - y\,dx$

(C) $dz = 2\,dy + dx$ (D) $dz = \frac{dy}{x} - \frac{y\,dx}{x^2}$

Solution

If "z" is a thermodynamic property that is a function of two independent variables (x,y), i.e., $z = z(x,y)$, then

$$dz = \left(\frac{\partial z}{\partial x}\right)_y dx + \left(\frac{\partial z}{\partial y}\right)_x dy = Mdx + Ndy \qquad \text{(i)}$$

Further if is a thermodynamic property, then from eq. (i)

$$\left(\frac{\partial M}{\partial y}\right)_x = \left(\frac{\partial N}{\partial x}\right)_y \qquad \text{(ii)}$$

(A) $M=y$, $N=x$

$$\text{Hence, } \left(\frac{\partial M}{\partial y}\right)_x = \left(\frac{\partial N}{\partial x}\right)_y = 1$$

Therefore, relation (A) identifies z as a thermodynamic property.

(B) $M = -y$, $N = x$

$$\text{Hence, } \left(\frac{\partial N}{\partial y}\right)_x = -1 \text{ and } \left(\frac{\partial N}{\partial x}\right)_y = -1, \text{ or} \left(\frac{\partial M}{\partial y}\right)_x \neq \left(\frac{\partial N}{\partial x}\right)_y$$

Therefore, relation (B) fails to identify z as a thermodynamic property.

(C) $M = 1, N = 2$

$$\text{Hence, } \left(\frac{\partial M}{\partial y}\right)_x = \left(\frac{\partial N}{\partial x}\right)_y = 0.$$

Therefore, relation (C) identifies z as a thermodynamic property.

(D) $M = -\frac{y}{x^2}, N = \frac{1}{x}$

$$\text{Hence, } \left(\frac{\partial M}{\partial y}\right)_x = \left(\frac{\partial N}{\partial x}\right)_y = -\frac{1}{x^2}.$$

Therefore, relation (D) identifies z as a thermodynamic property.

Hence, the answer is (B).

Q.2 Internal energy of a thermodynamic system is defined by the

(A) Zeroth law of thermodynamics
(B) First law of thermodynamics
(C) Second law of thermodynamics
(D) Third law of thermodynamics

Solution

Internal energy of a thermodynamic system is defined by the first law of thermodynamics.

Hence, the answer is (B).

Q.3 In a polytropic process described by PV^n = constant, if $n = 0$, the process is called as

(A) isobaric (B) isochoric
(C) isothermal (D) isentropic

Solution

For a polytropic process with $n = 0$, the process is defined by PV^0 = Constant, i.e., P = Constant This means the process is a constant pressure or isobaric process.

Hence, the answer is (A).

Q.4 The relation between the coefficient of performance of a refrigerator $(COP)_R$ and the coefficient of performance of a heat pump $(COP)_{HP}$ is

(A) $(COP)_{HP} = (COP)_R + 1$ (B) $(COP)_{HP} = (COP)_R - 1$
(C) $(COP)_{HP} = 1 - (COP)_R$ (D) $(COP)_{HP}\, X = (COP)_R = 1$

Solution

The COP of a heat pump is given by

$$(\text{COP})_{HP} = \frac{Q_1}{W} = \frac{Q_1}{Q_1 - Q_2}, \tag{i}$$

where Q_1 and Q_2 are the heat rejected to the hot body and heat absorbed from the cold body, respectively. The COP of a refrigerator is given by

$$(\text{COP})_R = \frac{Q_2}{W} = \frac{Q_1}{Q_1 - Q_2}, \qquad \text{(ii)}$$

From eq. (i) and eq. (ii), we can get

$$(\text{COP})_{HP} = (\text{COP})_R + 1$$

Hence, the answer is (A).

Q.5 If L_1, L_2 and L_3 are the latent heats of vaporization at the critical temperature of nitrogen, water and ammonia, respectively, then which one of the following, is true?

(A) $L_1 > L_2 > L_3$ (B) $L_1 > L_2$ and $L_2 = L_3$
(C) $L_1 < L_2 < L_3$ (D) $L_1 = L_2 = L_3$

Solution

At the critical point the liquid-vapour transition lines vanish for all the pure substances.

Hence, the answer is (D).

Q.6 A new temperature scale (°N) has been proposed where the normal freezing and normal boiling points of water are marked as 500 °N and 100 °N, respectively. If the temperature of a system is measured to be 0 °N, its temperature according to the Celsius scale (in °C) is ______

Solution

Let the new scale be of the form:

$$N = aC + b \qquad \text{(i)}$$

where N and C are the temperatures in the °N and °C temperature scales, respectively, and a and b are constants.

Now, at C=0, N=500 and at C=100, N=100. Hence from eq. (i)

$$a \times 0 + b = 500$$

$$b = 500 \qquad \text{(ii)}$$

And $a \times 100 + 500 = 100$

$$a = -4 \qquad \text{(iii)}$$

Hence from eq. (i) we can write the new temperature scale as

$$N = -4C + 500 \qquad \text{(iv)}$$

At $N = 0$ we get from eq. (iv):

$$-4C + 500 = 0$$

$$C = 125$$

Hence, the answer is 125.

Q.7 Let Z_1 represents the compressibility factor of air at 2 bar and 600 K, and Z_2 represents the compressibility factor of air at 1 bar and 300 K. If air is assumed to be an ideal gas having gas constant of 0.287 kJ/kg.K, then Z_1/Z_2 is __________.

Solution

The air is considered as an ideal gas for the two thermodynamic states. Hence, the compressibility factors Z_1 and Z_2 – both are unity.

$$\text{Therefore, } \frac{Z_1}{Z_2} = 1$$

Hence, the answer is 1.

Q.8 The rate of heat received by a heat engine from a source at 900 K is 600 kJ/s. The engine rejects heat to the sink of 300 K. The heat engine produces a power of 200 kW. The irreversibility rate (in kW) of the process is __________.

Solution

The rate of irreversibility is given by

$$\dot{I} = \dot{W}_{max} - \dot{W} \quad \text{(i)}$$

Now,

$$\dot{W}_{max} = \left(1 - \frac{T_2}{T_1}\right)\dot{Q}_1 \quad \text{(ii)}$$

$$\dot{W}_{max} = \left(1 - \frac{300}{900}\right) \times 600 \text{ kW}$$

$$\dot{W}_{max} = 400 \text{ kW}$$

Given that w=200 kW. Hence, from eq. (i) and eq. (ii)

$$I = (400\text{-}200) \text{ kW} = 200 \text{ kW}$$

Hence, the answer is 200

Q.9 An engine working on the air standard Diesel cycle has a compression ratio of 18. The cycle has a cut-off ratio of 1.7. If the ratio of specific heats of air is 1.4, then the thermal efficiency (in %) of the cycle (rounded off to 1 decimal place) is__________.

Solution

Given:

Compression ratio, $r_k = 18$.

Cut-off ratio, $r_c = 1.7$.

Ratio of specific heats, $\gamma = 1.4$.

Now the efficiency of an air standard Diesel cycle is given by

$$\eta_{Diesel} = 1 - \frac{1}{\gamma} \cdot \frac{1}{r_k^{\gamma-1}} \cdot \frac{r_c^{\gamma} - 1}{r_c - 1} = 64.6$$

Hence, the answer is 64.6

Q.10 - Q.22 carry two marks each.

Q.10 A system with rigid walls is initially at a temperature of T_1. It is used as the heat source for a heat engine, which rejects heat to a reservoir maintained at T_0 ($T_0 < T_1$). The specific heats of the system are constant. If the temperature of the system finally reduces to T_0, then the maximum work recoverable from the heat engine per unit mass of the system is

(A) $C_v \left[(T_1 - T_0) - T_0 \, In \left(\frac{T_1}{T_0} \right) \right]$ (B) $C_v (T_1 - T_0)$

(C) $C_v T_0 In \left(\frac{T_1}{T_0} \right)$ (D) $C_v \frac{T_1^2}{T_0}$

Solution

Let us refer to Fig. Q.10. The system with rigid walls transfers heat at constant volume to the heat engine. For evaluating the maximum work recoverable from the heat engine, we have to consider reversible processes in the heat engine till the constant volume system attains the dead state. Let the heat transferred from the system to the engine be Q_1. Since the heat transfer process also needs to be reversible and the temperature of the system decrease from T_1 to T_0 by an isochoric process, hence the process should result in increase of temperature of the working fluid of the heat engine from T_0 to T_1 through a constant volume process. Let us consider unit mass of both the system and working fluid in the heat engine.

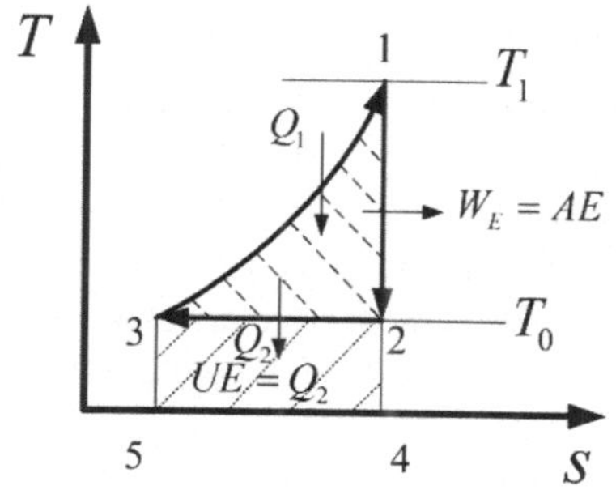

Fig. **Q.10**

Now, heat lost by the system = Heat gained by the working fluid in the heat engine = Q_1

$$\text{i.e., } (C_v)_g (T_1 - T_0) = (C_v)_{wf} (T_1 - T_0)$$

$$(C_v) = (C_v)_{wf} = C_{v'} \tag{i}$$

where $(C_v)_g$ and $(C_v)_{wf}$ are the specific heats at constant volume of the system and the working fluid of the heat engine, respectively.

Now the increase in entropy of the working fluid during the heat addition is given by

$$\Delta s_{wf} = s_4 - s_5 = \int_{T_0}^{T_1} \frac{C_v dT}{T} = C_v \ln\left(\frac{T_1}{T_0}\right). \tag{ii}$$

From Fig. Q10, The heat rejected by the working fluid reversibly at the temperature T_0 is given by the area 24532, i.e.

$$Q_2 = UE = T_0 \Delta s_{wf} = T_0 C_v \ln\left(\frac{T_1}{T_0}\right), \tag{iii}$$

where, UE is the unavailable energy. Hence, the maximum work recoverable from the heat engine is given by

$$\begin{aligned} W_{\max} &= AE = Q_1 - UE \\ &= C_v (T_1 - T_0) - T_0 C_v \ln\left(\frac{T_1}{T_0}\right) \\ &= C_v \left[(T_1 - T_0) - T_0 \ln\left(\frac{T_1}{T_0}\right) \right] \end{aligned}$$

Hence, the answer is (A).

Q.11 A reversible heat engine is operating between two reservoirs maintained at T_1 and T_2, where $T_1 > T_2$. Which one of the following is the most effective option for increasing its thermal efficiency?

(A) increasing T_1, while keeping T_2 constant
(B) decreasing T_1, while keeping T_2 constant
(C) increasing T_2, while keeping T_1 constant
(D) decreasing T_2, while keeping T_1 constant

Solution

From the second law of thermodynamics it can be deduced that all reversible heat engines operating between the same source and sink temperatures have the same efficiency and that is equal to the Carnot cycle efficiency.

$$\text{Hence, } \eta_{rev} = \eta_{Carnot} = 1 - \frac{T_2}{T_1} \tag{i}$$

From eq. (i), it can be seen that η_{rev} can be increased by either increasing T_1 or decreasing T_2. In order to conclude which one is more effective, i.e., whether increasing T_1 or decreasing T_2, let us consider once decreasing T_2 by an amount ΔT keeping T_1 constant and then by increasing T_1 by an amount ΔT keeping T_2

constant. Let us designate the corresponding efficiencies as (η_{rev}) and $(\eta_{rev})_2$, respectively.

$$\text{Now, } (\eta_{rev})_1 = 1 - \frac{T_2 - \Delta T}{T_1} = \frac{T_1 + \Delta T - T_2}{T_1} \quad \text{(ii)}$$

$$\text{And } (\eta_{rev})_2 = 1 - \frac{T_2}{T_1 + \Delta T} = \frac{T_1 + \Delta T - T_2}{T_1 + \Delta T} \quad \text{(iii)}$$

$$\text{Therefore, } \frac{(\eta_{rev})_1}{(\eta_{rev})_2} = \frac{T_1 + \Delta T}{T_1}$$

$$\frac{(\eta_{rev})_1}{(\eta_{rev})_2} = 1 + \frac{\Delta T}{T_1} > 1 \quad \text{(iv)}$$

Eq. (iv) shows that $(\eta_{rev})_1 > (\eta_{rev})_2$.

Hence, the answer is (D).

Q.12 A 4-m³ reservoir contains 10 kg of a real gas at 200 K. If this gas follows the van der Waal's equation of state with a = 0.0687 m⁶.kPa/kg², b = 0.00657 m³/kg and R = 0.187 kJ/kg.K, then the reservoir pressure (in kPa) is

(A) 93.5 (B) 94.6

(C) 95.7 (D) 101.3

Solution

van der Waal's equation of state is

$$\left(P + \frac{a}{v^2}\right)(v - b) = RT \quad \text{(i)}$$

$$\text{Now, } v = \frac{V}{m} = \frac{4}{10}\,m^3/\text{kg} = 0.4\ \text{m}^3/\text{kg}.$$

Therefore, from eq. (i)

$$\left\{P + \frac{0.0687}{(0.4)^2}\right\}(0.4 - 0.00657) = 0.187 \times 200$$

$$P = 94.6\ \text{kPa}$$

Hence, the answer is (B).

Q.13 Air at a pressure of 86 kPa and specific volume of 1 m³/kg is heated at constant pressure till it reaches 627 C. Air is assumed to be an ideal gas with constant specific heats. It has the gas constant of 0.287 kJ/kg.K and ratio of specific heats of 1.4. The change in specific entropy of air (in kJ/kg.K) during this process will be

(A) 1.104 (B) 0.740

(C) 0.788 (D) 0.529

Solution

Let the initial temperature be T_1. Then

$$T_1 = \frac{P_1 v_1}{R} = \frac{86 \times 1}{0.287} \text{ K} = 299.65 \text{ K}$$

The specific heat at constant pressure is obtained as

$$C_p = \frac{\gamma R}{\gamma - 1} = \frac{1.4 \times 0.287}{1.4 - 1} \text{kJ/kg. K} = 1.0045 \text{kJ/kg.K}$$

Now, the entropy change in a constant pressure process of an ideal gas is given by

$$\Delta s = C_p \ln\left(\frac{T_2}{T_1}\right) = 1.0045 \times \ln\left(\frac{627 + 273}{299.65}\right) \text{kJ/kg. K} = 1.104 \text{ kJ/kg. K}$$

Hence, the answer is (A).

Q.14 An air standard Otto cycle has compression ratio of 4. The compression ratio of this cycle is changed to 6. If the ratio of specific heats is 1.4, the percentage increase in its thermal efficiency will be

(A) 20.2 (B) 27.2

(C) 42.6 (D) 51.2

Solution

The efficiency of an Air-standard Otto cycle is given by

$$\eta_{Otto} = 1 - \frac{1}{r_k^{\gamma - 1}}$$

For $r_k = 4$, the cycle efficiency is

$$(\eta_{Otto})_1 = 1 - \frac{1}{4^{1.4-1}} = 0.4257$$

For $r_k = 6$, the cycle efficiency is

$$(\eta_{Otto})_2 = 1 - \frac{1}{6^{1.4-1}} = 0.5116$$

Hence, the percentage increase in efficiency is

$$\frac{(\eta_{Otto})_2 - (\eta_{Otto})_1}{(\eta_{Otto})_1} \times 100\% = \frac{0.5116 - 04.257}{0.4257} \times 100\% = 20.2\%$$

Hence, the answer is (A).

Q.15 In a mixture of gas there are 0.1 kmol of oxygen (O_2), 0.1 kmol of nitrogen (N_2) and 0.8 kmol of methane (CH_4). If the molar mass of O_2, N_2 and CH_4 are 32 kg/kmol, 28 kg/kmol and 16 kg/kmol, respectively, then the mass fraction of N_2 in the gas mixture is

(A) 0.100 (B) 0.170

(C) 0.148 (D) 0.680

Solution

Mass of O_2 present, $(m)_{O_2} = (n)_{O_2}(\mathcal{M})_{O_2} = 0.1 \times 32 \text{ kg} = 3.2 \text{ kg}$

Mass of N_2 present, $(m)_{N_2} = (n)_{N_2}(\mathcal{M})_{N_2} = 0.1 \times 28 \text{ kg} = 2.8 \text{ kg}$

Mass of CH_4 present, $(m)_{CH_4} = (n)_{CH_4}(\mathcal{M})_{CH_4} = 0.8 \times 16 \text{ kg} = 12.8 \text{ kg}$

Therefore, the mass fraction of N_2 in the mixture

$$= \frac{(m)_{N_2}}{(m)_{O_2} + (m)_{N_2} + (m)_{CH_4}} = 0.148$$

Hence, the answer is (C).

Q.16 A particular gas sample is initially maintained at 6000 cm³ and 100 kPa. It is compressed during a quasistatic process following the relation PV^2 = constant. The compression continues till the volume becomes 2000 cm³. The magnitude of the corresponding work transfer (in kj) (rounded off to 2 decimal places) is __________.

Solution

Given:

$P_1 = 100$ kPa

$V_1 = 6000 \text{ cm}^3 = 0.006 \text{ m}^3$

$V_2 = 6000 \text{ cm}^3 = 0.002 \text{ m}^3$

The compression process follows a polytropic relation: PV^2. This means that the polytropic index $n = 2$.

Let the pressure after compression be P_2.

$$\text{Therefor, } P_2 = P_1\left(\frac{V_1}{V_2}\right)^2 = 900 \text{ kPa}$$

The work supplied in the polytropic compression process is given by

$$W = \frac{P_2V_2 - P_1V_2}{n-1} = \frac{900 \times 0.002 - 100 \times 0.006}{2-1} \text{kJ} = 1.2 \text{ kJ}$$

Hence, the answer is 1.2.

Q.17 Carbon dioxide (CO_2) enters an adiabatic rigid nozzle steadily at 1 MPa and 500 C with a mass flow rate of 1.5 kg/s. The inlet area of the nozzle is 40 cm^2 and the exit velocity is 10 times of that at the inlet. If CO_2 can be considered as an ideal gas with gas constant of 0.19 kJ/kg.K. and the ratio of specific heats of 1.29, the exit temperature (in K) (rounded off to 1 decimal place) is ___________.

Solution

The density of CO_2 at the inlet of the nozzle,

$$\rho_1 = \frac{P_1}{RT_1} = \frac{10^6}{190 \times 773} \text{ kg/m}^3 = 6.81 \text{ kg/m}^3$$

Let the velocity at the inlet of the nozzle be V_1. Therefore,

$$\rho_1 A_1 V_1 = \dot{m}$$

$$V_1 = \frac{\dot{m}}{\rho_1 A_1} = \frac{1.5}{6.81 \times 0.004} \text{m/s} = 55.07 \text{m/s.}$$

Therefore, the exit velocity, $V_2 = 10V_1 = 550.7$ m/s.

Now, for an ideal gas $C_p = \frac{\gamma R}{\gamma - 1} = \frac{1.29 \times 190}{1.29 - 1} \frac{\text{kJ/kg.K}}{\text{kg}} = 845.17 \text{ kJ/kg. K}$

Noting the fact that for an ideal gas $h = C_p T$, and applying the steady flow energy equation across the nozzle one obtains

$$h_1 + \frac{V_1^2}{2} = h_2 + \frac{V_2^2}{2}$$

$$C_pT_1 + \frac{V_1^2}{2} = C_pT_2 + \frac{V_2^2}{2}$$

$$T_2 = T_1 + \frac{V_1^2 - V_2^2}{2C_p}$$

$$T_2 = \left\{773 + \frac{(55.07)^2 - (550.7)^2}{2 \times 845.17}\right\} \text{K} = 595.4 \text{ K}$$

Hence, the answer is 595.4

Q.18 A closed system containing 8 kg of gas undergoes an expansion process following the relation $PV^{1.2}$ = constant. The initial and final pressures are 1 MPa and 5 kPa, respectively, while the initial volume is 1 m³. If the specific internal energy of the gas decreases by 40 kJ/kg during the process, the heat transfer (in kJ) associated with the process (rounded off to 1 decimal place) is __________.

Solution

$P_1V_1^{1.2} = P_2V_2^{1.2}$ [i.e., the polytropic index $n = 1.2$]

$$V_2 = V_1\left(\frac{P_1}{P_2}\right)^{\frac{1}{1.2}} = 1 \times \left(\frac{1000}{5}\right)^{\frac{1}{1.2}} \text{m}^3 = 82.7\text{m}^3$$

Now, the work done by the system is obtained as

$$W = \frac{P_1V_1 - P_2V_2}{n-1} = \frac{1000 \times 1 - 5 \times 82.7}{1.2 - 1} \text{kJ} = 2932.5 \text{ kJ}$$

The total change in internal energy of the system

$\Delta U =$ The mass of the system × The change in specific internal energy

$$\Delta U = 8 \times (-40) \text{ kJ} = -320 \text{ kJ}$$

Now the first law of thermodynamics gives

$$Q = \Delta U + W$$

$$Q = (-320 + 2932.5) \text{ kJ} = 2612.5 \text{ kJ}$$

Hence, the answer is 2612.5.

Q.19 Saturation pressure of water at 5 C is 0.8725 kPa. If the latent heat of vaporization is 2489.1 kJ/kg and gas constant is 0.4615 kJ/kg.K, then the saturation pressure at 10 C (in kPa) (rounded off to 2 decimal places) is __________.

Solution

The Clausius-Clapeyron equation gives

$$\ln\left(\frac{P_2}{P_1}\right) = \frac{\Delta H_{Vap}}{R}\left(\frac{1}{T_1} - \frac{1}{T_2}\right)$$

$$\ln\left(\frac{P_2}{0.8725}\right) = \frac{2489.1}{0.4615}\left(\frac{1}{278} - \frac{1}{283}\right)$$

$$\ln\left(\frac{P_2}{0.8725}\right) = 03428$$

$$P_2 = 0.8725 \times e^{0.3428} \text{kPa} = 1.23 \text{ kPa}$$

Hence, the answer is 1.23.

Q.20 The turbine inlet conditions of a Rankine cycle are 10 MPa and 500 C, while the condenser pressure is 10 kPa, The enthalpy and entropy of saturated liquid at 10 kPa are 191.8 kJ/kg and 0.6492 kJ/kg.K, respectively, while the enthalpy and entropy of vapourizalion at 10 kPa are 2392.1 kJ/kg and 7.4996 kJ/kg.K, respectively. The enthalpy and entropy at the inlet to the turbine are 3375.1 kJ/kg and 6.5995 kJ/kg.K, respectively. The condenser outlet has saturated liquid. Neglecting the pump work, the thermal efficiency (in %) of the cycle (rounded off to 1 decimal place) is ____________.

Solution

Given:

Enthalpy at the inlet of the turbine, h_1 = 3375.1 kJ/K

Entropy at the inlet of the turbine, s_1 = 6.5995 kJ/kg.K

Enthalpy of saturated liquid at the condenser pressure, $h_f|_{Cond} = 191.8$ kJ/K

Entropy of saturated liquid at the condenser pressure, $s_f|_{Cond} = 0.6492$ kJ/kg. K

Enthalpy of saturated liquid at the condenser pressure, $h_g|_{Cond} = 2392.1$ kJ/K

Entropy of saturated liquid at the condenser pressure, $s_g|_{Cond} = 7.4996$ kJ/kg. K

Let the entropy at the outlet of the turbine be s_2.

Let the dryness fraction of steam at the condenser pressure be x_2.

Since the expansion process in the turbine is is-entropic, hence

$$s_2 = s_1$$

$$s_f|_{Cond} + x_2 \times \left(s_g|_{Cond} - s_f|_{Cond}\right) = s_1$$

$$x_2 = \frac{s_1 - s_f|_{Cond}}{s_g|_{Cond} - s_f|_{Cond}}$$

$$x_2 = \frac{6.5995 - 0.6492}{7.4996 - 0.6492} = 0.87$$

Hence, the enthalpy at the outlet of the turbine or the inlet to the condenser

$$h_2 = h_f|_{Cond} + x_2(h_g|_{Cond} - h_f|_{Cond})$$

$$= 191.8 + 0.87 \times (2392.1 - 191.8) \text{kJ/kg}$$

$$= 2106.06 \text{ kJ/kg}$$

Therefore, heat rejected in the condenser, $Q_2 = h_2 - h_f|_{Cond} = 1914.26$ kJ/kg

And, the work done by the turbine, $W_T = h_1 - h_2 = 1269.04$ kJ/kg

Now, neglecting the pump work the heat added can be written as

$$Q_1 = Q_2 + W_T = 3183.3 \text{ kJ/kg}$$

Therefore, the thermal efficiency of the cycle,

$$\eta_{th} = \frac{W_T}{Q_1} \times 100\% = 39.9\%$$

Hence, the answer is 39.9.

Q.21 The minimum and maximum temperatures of an air standard Brayton cycle are 300 K and 1100 K, respectively. The pressure ratio of this cycle is 6. The ratio of specific heats is 1.4 and the specific heats are constant. For this cycle, the ratio of network output to the turbine work (rounded off to 2 decimal places) is________________.

Solution

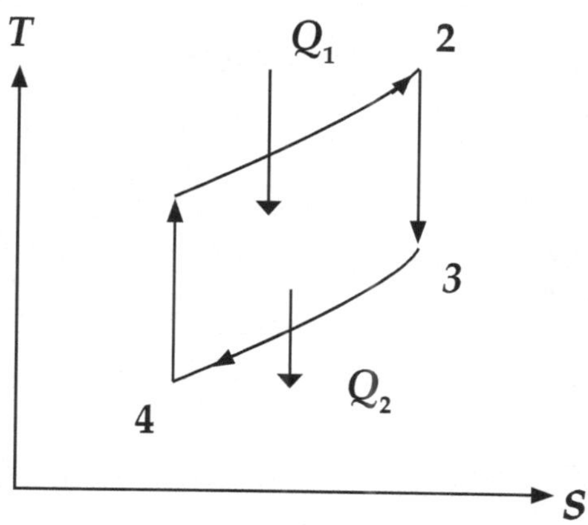

Fig. Q.21

Given:

$$T_4 = 300 \text{ K}$$

$$T_2 = 1100 \text{ K}$$

$$\frac{P_1}{P_4} = \frac{P_2}{P_3} = 6$$

$$P_1 = P_2$$

$$P_3 = P_4$$

Therefore,

$$T_1 = T_4\left(\frac{P_1}{P_4}\right)^{\frac{\gamma-1}{\gamma}} = 300\times(6)\frac{1.4-1}{1.4}\text{K} = 500.55\text{K}$$

$$T_3 = T_2\left(\frac{P_3}{P_2}\right)^{\frac{\gamma-1}{\gamma}} = 1100\times\left(\frac{1}{6}\right)^{\frac{1.4-1}{1.4}}\ \text{K} = 659.27\ \text{K}$$

Now,

The specific turbine work,

$$w_T = C_p\,(T_2 - T_3) = C_p\,(1100 - 659.27)\ \text{kJ/kg} = 440.73C_p\ \text{kJ/kg}$$

And the specific compressor work

$$w_C = C_p\,(T_1 - T_4) = C_p\,(500.55 - 300.00)\ \text{kJ/kg} = 200.55C_p\ \text{kJ/kg}$$

Therefore, the net specific work output of the cycle,

$$w_{net} = w_T - w_c = 240.18C_p\ \text{kJ/kg}$$

The ratio of net work output of the cycle to the turbine work

$$= \frac{240.18}{440.73} = 0.55$$

Hence, the answer is 0.55.

Q.22 The specific humidity of air at 100 kPa is 0.015 kg of vapour per kg of dry air. The partial pressure of vapour (in kPa) in the existing state (rounded off to 2 decimal places) is ___________.

Solution

Given:

Specific humidity, $\omega = 0.015$ kg of water vapour per kg of dry air

Atmospheric pressure, $P_a = 100$ kPa

Let the partial pressure of water vapour in the air be be P_v kPa.

Now, $\omega = 0.622\dfrac{P_v}{P_a - P_v}$

$$0.622\frac{P_v}{100 - P_v} = 0.015$$

$P_v = 2.35$ kPa

Hence, the answer is 2.35.

GATE 2019
THERMODYNAMICS

Q. 1 – Q. 9 carry one mark each & Q. 10 – Q. 22 carry two marks each.

Notations used:

c_p: Specific heat at constant pressure; c_v: Specific heat at constant volume; h: Specific enthalpy; s: Specific entropy; v: Specific volume; P: Pressure; V: Volume; T: Temperature; R: Gas constant; R_u : Universal Gas constant

Q.1 One kg of an ideal gas (molecular weight = X) occupies a certain volume V at temperature T and pressure. Four kgs of another ideal gas (molecular weight = $2X$) is added to the first gas keeping the volume V and temperature T same as before. The final pressure is

(A) $2P_1$ (B) $3P_1$

(C) $4P_1$ (D) $5P_1$

Solution

The number of moles of the gas with molecular weight X, $n_1 = \dfrac{1}{x}$

The number of moles of the gas with molecular weight 2X, $n_2 = \dfrac{4}{2X} = \dfrac{2}{X}$

Hence, the mole fraction of the gas with molecular weight, X, $x_1 = \dfrac{n_1}{n_1 + n_2} = \dfrac{1}{3}$

And the mole fraction of the gas with molecular weight, $2X$, $x_2 = \dfrac{n_2}{n_1 + n_2} = \dfrac{2}{3}$

Now the partial pressure of a gas is proportional to its mole fraction. Let the fial pressure be P. Now, for the given volume and temperature the given mass of gas with molecular weight X exerts a partial pressure of P_1.

Therefore, $P_1 = x_1 P$

$$P = \frac{P_1}{x_1} = \frac{P_1}{\frac{1}{3}} = 3P_1$$

Hence, the answer is (B).

Q.2 Which of the following is an intensive thermodynamic property?

(A) Enthalpy
(B) Internal energy
(C) Entropy
(D) Pressure

Solution

Pressure is an intensive thermodynamic property.

Hence, the answer is (D).

Q.3 A thermodynamic process for a substance is represented as a constant pressure process on a *P-V* diagram, and a constant temperature process on a *T-V* diagram. Which of the following statements is TRUE?

(A) The substance is an ideal gas.
(B) The substance is a van der Waals gas.
(C) The substance undergoes a phase change.
(D) Such a process is not possible.

Solution

During a phase-change process both the temperature and pressure remain constant, while the volume changes.

Hence, the answer is (C).

Q.4 Addition of a reheat process to a simple Rankine cycle will always

(A) increase both efficiency and work output of the cycle.
(B) increase efficiency but not necessarily work output of the cycle.
(C) increase work output but not necessarily efficiency of the cycle.
(D) decrease both efficiency and work output of the cycle.

Solution

Addition of a reheat process to a simple Rankine cycle increases the work output, but at the same time the total heat input to the cycle also increases. Hence the work output of the cycle increases, but the efficiency will not increase necessarily.

Hence, the answer is (C).

Q.5 A Carnot cycle implemented using an ideal gas in a piston-cylinder system will have

(A) Work transfer in two parts of the cycle and heat transfer in the other two parts of the cycle.
(B) Work transfer in all four parts of the cycle and heat transfer in two parts of the cycle.
(C) Work transfer in two parts of the cycle and heat transfer in all four parts of the cycle.
(D) Work transfer in all four parts of the cycle and heat transfer in all four parts of the cycle.

Solution

A Carnot cycle consists of two reversible adiabatic processes and two reversible isothermal processes. During the reversible adiabatic processes work is transferred adiabatically. Work transfer also takes place during the isothermal heat addition and heat rejection processes.

Hence, the answer is (B).

Q.6 The "degrees of freedom" for a pure substance at its triple point is

(A) 3 (B) 2

(C) 1 (D) 0

Solution

For a thermodynamic equilibrium, the phase rule gives

$$F = C - P + 2,$$

where F is the number of degrees of freedom, C is the number of components and P is the number of phases. For a a pure substance at triple point, $C = 1$ and $P = 3$.

Therefore, $F = 1 - 3 + 2 = 0$.

Hence, the answer is (D).

Q.7 The differential of the Gibbs function (g) for a simple compressible system can be represented as $dg = v\,dP - s\,dT$. Using the appropriate Maxwell relation derived from the above equation, the quantity, $\left[-\left(\frac{\partial S}{\partial P}\right)_T\right]$ for an ideal gas is

(A) $\frac{R}{T}$ (B) $\frac{R}{P}$

(C) $\frac{P}{R}$ (D) $\frac{T}{R}$

Solution

$$dg = vdP - sdT$$

Therefore, $\left[-\left(\frac{\partial s}{\partial P}\right)_T\right] = \left[\left(\frac{\partial v}{\partial T}\right)_P\right]$

$$= \left\{\frac{\partial\left(\frac{RT}{P}\right)}{\partial T}\right\}_P$$

$$= \frac{R}{P}$$

Hence, the answer is (B).

Q.8 A Carnot engine receives 2000 kJ heat from a source at 1200 K and rejects 500 kJ heat to sink. The sink temperature (K) is ________.

Solution

$$\frac{T_2}{T_1} = \frac{Q_2}{Q_1}$$

$$T_2 = \left(\frac{\dot{Q}_2}{Q_1}\right) T_1$$

$$= \left(\frac{500}{2000}\right) \times 1200\text{K}$$

$$= 300 \text{ K}$$

Hence, the answer is 300 .

Q.9 A closed vessel contains a pure substance with 0.1 m³ of saturated liquid and 0.9 m³ of saturated vapor. The specific volume of the liquid and vapor phases are 0.000843 m³/kg and 0.02671 m³/kg respectively. The percentage of vapor in the mixture on mass basis (rounded off to two decimal places) is ________.

Solution

Mass of the vapour, $m_g = \frac{V_g}{v_g} = \frac{0.9}{0.02671} \text{ kg} = 33.7 \text{ kg}$

Mass of the liquid, $m_f = \frac{V_f}{v_f} = \frac{0.1}{0.000843} \text{ kg} = 118.62 \text{ kg}$

Therefore, the percentage of vapour by mass in the mixture

$$= \frac{33.7}{33.7 + 118.62} \times 100\% = 22.12\%$$

Hence, the answer is 22.12.

Q.10 A gas obeys the van der Waals equation of state $\left(P + \frac{a}{v^2}\right)(v - b) = RT$, where a and b are van der Waals constants. The compressibility factor $\left(Z = \frac{Pv}{RT}\right)$ in the limit of high specific volume ($b \ll v$) is

(A) $1 + \frac{1}{v}\left(b - \frac{a}{RT}\right)$ (B) $1 + \frac{1}{v}\left(b + \frac{a}{RT}\right)$

(C) $1 - \frac{1}{v}\left(b + \frac{a}{RT}\right)$ (D) $1 - \frac{1}{v}\left(b - \frac{a}{RT}\right)$

Solution to Q.10

$$\left(P+\frac{a}{v^2}\right)(v-b)=RT$$

$$P=\frac{RT}{v-b}-\frac{a}{v^2}$$

Therefore, the compressibility factor can be obtained as

$$Z=\frac{Pv}{RT}$$

$$=\frac{v}{RT}\left(\frac{RT}{v-b}-\frac{a}{v^2}\right)$$

$$=\frac{v}{v-b}-\frac{a}{RvT}$$

$$=\frac{1}{1-(b/v)}-\frac{a}{RvT}$$

$$=1+\frac{b}{v}-\frac{a}{RvT}\left[\text{Since}\frac{b}{v}\approx 0\right]$$

$$=1+\frac{1}{v}\left(b-\frac{a}{RT}\right)$$

Hence, the answer is (A).

Q.11 Consider the thermodynamic relation, $\left(\frac{\partial C_p}{\partial P}\right)_T=-T\left(\frac{\partial^2 v}{\partial T^2}\right)_P$. For a certain gas, if the isobaric thermal expansion coefficient, $\alpha=\frac{1}{T}$ (where T is the temperature measured in K), then the value of $\left(\frac{\partial C_p}{\partial P}\right)_T$ is,

(A) 0

(B) 1

(C) 2

(D) $\frac{v}{T}$

Solution to Q.11

$$\left(\frac{\partial C_p}{\partial P}\right)_T=-T\left(\frac{\partial^2 v}{\partial T^2}\right)_P$$

$$\alpha=\frac{1}{T}$$

$$\frac{1}{v}\left(\frac{\partial v}{\partial T}\right)_P=\frac{1}{T}$$

$$\left(\frac{\partial v}{\partial T}\right)_P=\frac{v}{T}$$

$$\text{Now, } \left(\frac{\partial C_p}{\partial P}\right)_T = -T\left(\frac{\partial^2 v}{\partial T^2}\right)_P$$

$$= -T\left\{\frac{\partial}{\partial T}\left(\frac{\partial v}{\partial T}\right)_P\right\}_P$$

$$= -T\frac{\partial}{\partial T}\left(\frac{v}{T}\right)_P$$

$$= -T\frac{\partial}{\partial T}\left(\frac{R}{p}\right)_P$$

$$= 0$$

Hence, the answer is (A).

Q.12 A piston-cylinder system has an initial volume of 0.1 m³ and contains an ideal gas (= 0.74 kJ/kg.K, R = 0.288 kJ/kg.K) at 1.5 bar and 298 K. The piston is moved to compress the gas until the pressure and temperature reach 10 bar and 423 K respectively. During this process, 20 kJ of work is done on the gas. The magnitude of heat transfer (kJ) during this process is

(A) 134.8 (B) 36.2

(C) 8.2 (D) 3.8

Solution

Given:

$P_1 = 1.5 \text{ bar} = 1.5 \times 10^5 \text{ N/m}^2$

$T_1 = 298$ K

$V_1 = 0.1 \text{ m}^3$

$T_2 = 423$ K

$C_v = 0.74$ kJ/kg. K

R = 0.288 kJ/kg.K

W = –20 kJ

If m is the mass of the gas then

$PV = mRT_1$

$$m = \frac{P_1V_1}{RT_1}$$

$$= \frac{1.5\times 10^5 \times 0.1}{288\times 298} \text{ kg}$$

$$= 0.175 \text{ kg}$$

From the first law of thermodynamics for a closed system:

$Q = \Delta U + W$

$Q = mC_v\,(T_2 - T_1) + W$

$= \{0.175 \times 0.288 \times (423\text{-}298) - 20\}$ kJ

$= -3.8$ kJ

Hence, the answer is (D).

Q.13 A reversible cyclic device absorbs 400 kJ/s heat from a reservoir maintained at 265 K and rejects heat to another reservoir at 298 K. The magnitude of the rate of work done (kW) by the device is

(A) 849.81 (B) 449.81

(C) 355.70 (D) 49.81

Solution

$$W = \left(1 - \frac{T_2}{T_1}\right) Q_1$$

$$= \left(1 - \frac{298}{265}\right) \times 400 \text{ kJ}$$

$$= -49.81 \text{ kJ}$$

Hence, the answer is (D).

Q.14 In a reversible process, an ideal gas (c_p = 1.04 kJ/kg.K, R = 0.297 kJ/kg.K) at 293 K, is compressed in a cylinder from 100 kPa to 500 kPa. During this compression process, the relation between pressure and volume is expressed as, $Pv^{1.3}$ = constant.

In the following options, mark the closest value for the entropy change per kilogram (J/kg.K).

(A) 91 (B) 91

(C) 864 (D) 864

Solution

$$\frac{T_2}{T_1} = \left(\frac{p_2}{p_1}\right)^{\frac{n-1}{n}}$$

$$= \left(\frac{500}{100}\right)^{\frac{1.3-1}{1.3}}$$

$$= 1.45$$

$$\therefore \Delta s = C_p \ln\left(\frac{T_2}{T_1}\right) - R \ln\left(\frac{p_2}{p_1}\right)$$

$$= [1040 \times \ln(1.45) - 297 \times \ln(5)] \text{ J/kg.K}$$

$$= -91 \text{ J/kg.K}$$

Hence, the answer is (A).

Q.15 The sublimation pressure of water vapor at 233 K is 13 Pa. Assume the water vapor to behave as an ideal gas (R = 0.46 kJ/kg.K), and the latent heat of sublimation is 2840 kJ/kg. The sublimation pressure of water vapor at 213 K, (rounded off to 2 decimal places), is __________ Pa.

Solution

From the Clausius-Clapeyron equation,

$$\ln\left\{\frac{(P_{sub})_1}{(P_{sub})_2}\right\} = \left(\frac{\Delta H_{sub}}{R}\right)\left[\frac{1}{T_2} - \frac{1}{T_2}\right]$$

$$\ln\left\{\frac{13}{(P_{sub})_2}\right\} = \left(\frac{2840}{0.46}\right)\left[\frac{1}{213} - \frac{1}{233}\right]$$

$$\ln\left\{\frac{13}{(P_{sub})_2}\right\} = 2.488$$

$$(P_{sub})_2 = \frac{13}{e^{2.488}} = 1.08 \text{ Pa}$$

Hence, the answer is 1.08.

Q.16 In a heating and humidification process, dry air (*d.a*) with relative humidity of 10% at ambient temperature of 10°C, is mixed with superheated steam at temperature 'T'. The resultant mixture is at 25°C with relative humidity of 50%. The total pressure before and after the process remains the same at 102 kPa. The saturated pressure of water vapor (*v*) at 10°C and 25°C are 1.2281 kPa, and 3.1698 kPa, respectively. The specific humidity in grams of water vapour per kg of dry air, can be calculated using w = 622 $\frac{p_v}{p_{d.a}}$, where p_v and $p_{d.a}$ are the partial pressures of water vapour and dry air respectively. The quantity of steam in grams (rounded off to 2 decimal places) added per kg of dry air is __________.

Solution

Given:

Relative humidity at inlet, $\varphi_1 = \dfrac{(p_v)_1}{(p_{sat})_1} = 0.1$

Dry bulb temperature at inlet, $t_1 = 10°$

Relative humidity at inlet, $\varphi_2 = \dfrac{(p_v)_2}{(p_{sat})_2} = 0.5$

Dry bulb temperature at inlet, $t_2 = 25°C$

Therefore, $(p_v)_1 = 0.1\,(p_{sat})_1 = 0.1 \times 1.2281 \text{ kPa} = 0.12281 \text{ kPa}$

Now, the specific humidity at inlet, $\omega_1 = 0.622\dfrac{(p_v)_1}{p_{atm} - (p_v)_1} = 0.00075$

The specific humidity at outlet, $\omega_2 = 0.622\dfrac{(p_v)_2}{p_{atm} - (p_v)_2} = 0.00982$

Therefore, the amount of steam added per kg of dry air

$= \omega_2 - \omega_1 = (0.00982\text{-}0.00075)$ kg = 9.06 grams

Hence, the answer is 9.06.

Q.17 In a vapor compression system, a refrigerant leaves the evaporator and enters the compressor in saturated vapor condition at 0 C. The specific enthalpies of the saturated liquid and the saturated vapor at 0 C are 50 kJ/kg and 250 kJ/kg, respectively. The refrigerant leaves the compressor with a specific enthalpy of 300 kJ/kg. From the exit of the condenser, the refrigerant is throttled to the evaporator pressure. If the coefficient of performance (COP) of the refrigerator is 2.8 then the dryness fraction of the refrigerant entering the evaporator is __________.

Solution

Given:

$$h_f|_{0°C} = 50 \text{ kJ/kg}$$

$$h_g|_{0°C} = 250 \text{ kJ/kg}$$

$$h_{exit}|_{Comp} = 300 \text{ kJ/kg}$$

$$\text{COP=2.8}$$

Hence $W|_{Comp} = h_{exit}|\text{-}_{Comp} - h_g|_{0°C}$ =50 kJ/kg

$$\text{Now, } COP = \frac{h_g|_{0°C} - h_{inlet}|_{Evap}}{W|_{Comp}}$$

$$\frac{250 - h_{inlet}|_{Evap}}{50} = 2.8$$

$$h_{inlet}|_{Evap} = 110 \text{ kJ/kg}$$

Let the dryness fraction of the refrigerat vapour at the evaporator inlet by X. Then

$$h_{inlet}|_{Evap} = h_f|_{(0°C)} + x\,(h_g|_{(0°C)} - h_f|_{(0°C)})$$

$$x = \frac{hinlet|_{Evap} - hf|_{0°C}}{hg|_{0°C} - hf|_{0°C}} = 0.3$$

Hence, the answer is 0.3.

Q.18 Hot air, assumed as an ideal gas (c_p = 1000 J/kg.K, γ = 1.4) enters a gas turbine at 10 bar, 1000 K and leaves at a pressure of 5 bar. Subsequently it expands in a nozzle to a pressure of 1 bar. Assume both these processes to be reversible and adiabatic. If the inlet velocity of the air to the nozzle is negligible, the final velocity (m/s, rounded off to 1 decimal place) of air at the exit of the nozzle is ________.

Solution

Given:

Pressure at the inlet of the gas turbine, P_1 = 10 bar

Temperature at the inlet of the gas turbine, T_1 = 1000 K

Pressure at the outlet of the gas turbine=Pressure at the inlet of the nozzle, P_2 = 5 bar

Pressure at the outlet of the nozzle, P_3 = 1 bar

The specific heat at constant pressure of the hot air, C_p = 1000 J/kg.K

Adiabatic index, $\gamma = 1.4$

The velocity at the inlet of the nozzle is negligible, i.e. $V_2 = 0$

Let the temperature at outlet of the gas turbine, i.e. at the inlet of the nozzle be T_2 and that at the outlet of the nozzle be T_3.

Since the expansion processes in both the turbine and nozzle are reversible adiabatic and working fluid is an ideal gas, hence

$$\frac{T_2}{T_1} = \left(\frac{P_2}{P_1}\right)^{\frac{\gamma-1}{\gamma}}$$

$$T_2 = T_1\left(\frac{P_2}{P_1}\right)^{\frac{\gamma-1}{\gamma}}$$

$$T_2 = 820.34 \text{ K}$$

Further,

$$\frac{T_3}{T_2} = \left(\frac{P_3}{P_2}\right)^{\frac{\gamma-1}{\gamma}}$$

$$T_3 = T_3\left(\frac{P_3}{P_2}\right)^{\frac{\gamma-1}{\gamma}}$$

$$T_3 = 517.95 \text{ K}$$

Now, applying the steady flow energy equation between the inlet and outlet of nozzle then,

$$h_2 + \frac{V_2^2}{2} = h_3 + \frac{V_3^2}{2}$$

$$\frac{V_3^2}{2} = h_2 - h_3 \text{ [Since } V_2 = 0]$$

$$V_2 = \sqrt{2(h_2 - h_3)}$$

$$V_2 = \sqrt{2C_p(T_2 - T_3)} \text{ [Since for a perfect gas } h = C_p T]$$

$= 777.67$ m/s

Hence, the answer is 777.67

Q.19

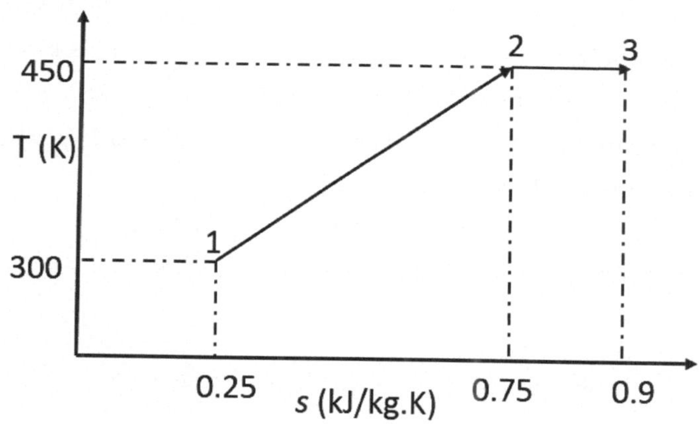

The "*T-s* diagram" for a thermodynamic process is shown in the figure. The heat transferred (kJ/kg) during the reversible process 1-3 is __________.

Solution

$$q = \int_1^3 Tds = \int_1^2 Tds + \int_2^3 Tds$$

$$= \frac{1}{2} \times (0.75 - 0.25) \times (450 - 300) + (0.75 - 0.25) \times 300 + (0.9 - 0.75) \times 450 \text{ kJ/kg}$$

$= 255$ kJ/kg

Hence, the answer is 255

Q.20 In a simple Rankine cycle, superheated steam enters the turbine at 100 bar, 500 C (h = 3375.1 kJ/kg, s = 6.5995 kJ/kg.K) at a rate of 1000 kg/s. It expands isentropically in the turbine to the condenser pressure. The temperature in the condenser is 30 C (h_f = 125.74 kJ/kg, s_f = 0.4368 kJ/kg.K, h_g = 2555.6 kJ/kg, s_g = 8.4520 kJ/kg.K). Saturated water from the condenser is pumped back to the boiler. Neglecting the pump work, the efficiency (in percentage, rounded off to 1 decimal) of the cycle is ________.

Solution

Given:

Enthalpy at the inlet of the turbine, h_1 = 3375.1 kJ/K

Entropy at the inlet of the turbine, s_1 = 6.5995 kJ/kg.K

Enthalpy of saturated liquid at the condenser pressure, $h_f|_{Cond}$ = 125.74 kJ/K

Entropy of saturated liquid at the condenser pressure, $s_f|_{Cond}$ = 0.4368 kJ/kg.K

Enthalpy of saturated vapour at the condenser pressure, $h_g|_{Cond}$ = 2555.6 kJ/kg.

Entropy of saturated vapour at the condenser pressure, $s_g|_{Cond}$ = 8.4520 kJ/kg.K

Let the entropy at the outlet of the turbine be s_2.

Let the dryness fraction of steam at the condenser pressure be x_2.

Since the expansion process in the turbine is isentropic, hence

$$s_2 = s_1$$

$$s_f\,|_{Cond} + x_2 \times (s_g\,|_{Cond} - s_f\,|_{Cond}) = s_1$$

$$x_2 = \frac{s_1 - s_f\,|_{Cond}}{s_g\,|_{Cond} - s_f\,|_{Cond}}$$

$$x_2 = \frac{6.5995 - 0.4368}{8.4520 - 0.4368} = 0.77$$

Hence, the enthalpy at the outlet of the turbine or the inlet to the condenser

$$h_2 = h_f\,|_{Cond} + x_2\left(h_g\,|_{Cond} - h_f\,|_{Cond}\right)$$

$$= 125.74 + 0.77 \times (2555.6 - 125.74)\ \text{kJ/kg}$$

$$= 1997.5\ \text{kJ/kg}$$

Therefore, heat rejected in the condenser, $Q_2 = h_2 - h_f\,|_{Cond} = 1871.76$ kJ/kg

And, the work done by the turbine, $W_T = h_1 - h_2 = 1377.6$ kJ/kg

Now, neglecting the pump work the heat added can be written as

$Q_1 = Q_2 + W_T$ = 3249.36 kJ/kg

Therefore, the thermal efficiency of the cycle,

$$\eta_{th} = \frac{W_T}{Q_1} \times 100\% = 42.39\%$$

Hence, the answer is 42.39.

Q.21 A rigid, adiabatic container has two parts (A) and (B) separated by a piston. Initially, part (A) is filled with 1 kg of an ideal gas (c_v = 750 J/kg.K, R = 500 J/kg.K) at 1 bar, 600 K. The part (B) is filled with 1 kg of the same ideal gas at 2 bar, 400 K. The piston is now removed. In the final state, the pressure (bar, rounded off to two decimal places) will be ______________.

Solution

The internal energy remains constant in a free-expansion process. If T is the temperature after mixing, then

$m_1 C_v T_1 + m_2 C_v T_2 = (m_1 + m_2) C_v T$

$$T = \frac{m_1 T_1 + m_2 T_2}{m_1 + m_2}$$

$$= \frac{1 \times 600 + 1 \times 400}{1+1} \text{K}$$

$$= 500 \text{ K}$$

The initial volume of part A, $V_1 = \frac{m_1 R T_1}{P_1} = \frac{1 \times 500 \times 600}{100000} \text{m}^3 = 3\text{m}^3$

The initial volume of part B, $V_2 = \frac{m_2 R T_2}{P_2} = \frac{1 \times 500 \times 400}{200000} \text{m}^3 = 1\text{m}^3$

Hence, the final volume after mixing, $V = V_1 + V_2 = 4 \text{ m}^3$

Let the final pressure after mixing be P. Now from the equation of state

$$P = \frac{(m_1 + m_1) R T}{V} = \frac{2 \times 500 \times 500}{4 \times 100000} \text{bar} = 1.25 \text{ bar}$$

Hence, the answer is 1.25 .

Q.22 In a steady flow process, superheated steam at 70 bar and 450 C, is throttled to 30 bar in an insulated valve, at a rate of 2 kg/s. The changes in the kinetic energy and potential energy during the process are assumed to be zero. The rate of entropy generation (kJ/s.K rounded off to three decimal places) during the throttling process is __________.

The thermodynamic property data for superheated steam at 70 bar and 450 C: h = 3288.3 kJ/kg, s = 6.6353 kJ/kg.K

The thermodynamic property data for superheated steam at 30 bar are given in the following table.

T(°C)	h(kJ/kg)	s(kJ/kg K)
400	3231.7	6.9235
450	3344.9	7.0856

Solution

Given:

$h_1 = 3288.3$ kJ/kg

$s_1 = 6.6353$ kJ/kg.K

Now for a throttling process with negligible changes in potential and kinetic energies, the enthalpy remains constant, i.e.,

$$h_1 = h_2$$

$$h_2 = 3288.3 \text{ kJ/kg}$$

Let the entropy after throttling be s_2. From the given data interpolation gives

$$\frac{s_2 - 6.9235}{7.0856 - 6.9235} = \frac{3288.3 - 3231.7}{3344.9 - 3231.7}$$

$$s_2 = 7.00455 \text{ kJ/kg. K}$$

Hence the rate of entropy generation,

$$\dot{S} = \dot{m}(s_2 - s_1) = 2 \times (7.00455 - 6.6353) \text{ kJ/s. K} = 0.74 \text{ kJ/s. K}$$

Hence, the answer is 0.74

GATE 2018
THERMODYNAMICS

Q1. When a fixed mass of air-water vapour mixture is heated at constant pressure,

(A) both relative and specific humidity decrease.

(B) relative humidity decreases, but specific humidity remains unchanged.

(C) specific humidity decreases, but relative humidity remains unchanged.

(D) both relative and specific humidity increase.

Solution:

When fixed mass of air-water vapor mixture heated at constant pressure then relative humidity decreases but specific humidity remains constant.

Hence, the answer is (B).

Q2. For a reversible isothermal expansion of one mole of an ideal gas from state 1 to state 2, the magnitude of work done is

(A) $RT\ln\left(\frac{P_1}{P_2}\right)$ (B) $P_2V_2 - P_1V_1$

(C) $R\ln\left(\frac{V_1}{V_2}\right)$ (D) 0

Solution:

For isothermal process,

T= constant, $\Delta T = 0$, $\Delta U = 0$

Also, for an ideal gas the equation of state is $\frac{RT}{V}$.

Now the work done for an infinitesimal expansion dV at a pressure P is given by $\delta W = PdV$

Hence the work done from state 1 to state 2 is given by

$$W = \int_{V_1}^{V_2} PdV$$

$$= \int_{V_1}^{V_2} \frac{RT}{V} dV$$

$$= RT\int_{V_1}^{V_2} \frac{dV}{V} \text{ [Since } T\text{=constant for an isothermal process]}$$

$$= RT\ln\left(\frac{V_2}{V_1}\right)$$

$$= RT\ln\left(\frac{P_1}{P_2}\right) \text{ [Since for an isothermal process of an ideal gas } \frac{V_2}{V_1} = \frac{P_1}{P_2}$$

Hence, the answer is (A).

Q3. The statement which is NOT a consequence of the first law of thermodynamics is

(A) Heat is a path function

(B) Energy is a property of a system

(C) Energy of an isolated system is not conserved

(D) A perpetual motion machine of the first kind is not possible

Solution:

The statement which is not consequence of 1st law of thermodynamic is "Energy of a system is not conserved", which is a false statement.

Hence, the answer is (C).

Q4. For a refrigerator absorbing heat Q_L from a cold region and rejecting heat Q_H to a hot region, the coefficient of performance is written as

(A) $\frac{Q_L}{Q_H - Q_L}$ (B) $\frac{Q_H}{Q_H - Q_L}$

(C) $\frac{Q_H - Q_L}{Q_L}$ (D) $\frac{Q_L}{Q_H}$

Solution:

COP of refrigerator, $\frac{Q_L}{Q_H - Q_L}$

Hence, the answer is (A).

Q5. The value of the compressibility factor at the critical point evaluated using the van der Waals equation of state is

(A) $\frac{2}{7}$ (B) $\frac{5}{8}$ (C) $\frac{3}{8}$ (D) $\frac{1}{7}$

Solution:

van der Waal's equation of state is

$$P = \frac{RT}{v-b} - \frac{a}{v^2} \quad \text{(i)}$$

Now at the critical temperature,

$$\left(\frac{\partial P}{\partial v}\right)_{T=T_C} = -\frac{RT_C}{(v_C - b)^2} + \frac{2a}{v_c^3} = 0 \quad \text{(ii)}$$

And

$$\left(\frac{\partial^2 P}{\partial v^2}\right)_{T=T_C} = \frac{2RT_c}{(v_c - b)^2} - \frac{6a}{v_c^4} = 0 \quad \text{(iii)}$$

Solving eqs. (ii) and (iii) simultaneously one obtains

$$b = \frac{1}{3} v_c \quad \text{(iv)}$$

$$R = \frac{8a}{9T_c v_c} \quad \text{(v)}$$

Substituting for b and R from eqs. (iv) and (v) in eq. (i) at the critical state is

$$P_c = \frac{8a}{9T_c v_c} \times \frac{T_c}{v_c - \frac{1}{3} v_c} - \frac{a}{v_c^2}$$

$$a = 3P_c v_c^2 \quad \text{(vi)}$$

Substituting eq. (vi) in eq. (v) gives

$$R = \frac{8}{3} \frac{P_c v_c}{T_c}$$

$$\therefore Z = \frac{P_c v_c}{RT_c} = \frac{3}{8} \quad \text{(vii)}$$

Hence, the answer is (C).

Q6. The vapour pressure of a liquid at 8°C is 2.7 kPa. Its enthalpy of vaporization is constant and equal to 42700 kJ/kmol. Take R = 8.314 kJ/kmol.K. The temperature (in °C) at a vapour pressure of 13.5 kPa is

(A) 58.7 (B) 51.4 (C) 44.3 (D) 35.2

Solution:

From the Clausius-Clapeyron equation:

$$\ln \frac{p_1}{p_2} = \frac{\Delta H_{vap}}{R} \left(\frac{1}{T_2} - \frac{1}{T_1} \right)$$

$$\ln \left(\frac{2.7}{13.5} \right) = \frac{42700}{8.314} \left(\frac{1}{T_2} - \frac{1}{281} \right)$$

$$\ln(0.2) = \frac{42700}{8.314}\left(\frac{1}{T_2} - \frac{1}{281}\right)$$

$$\frac{-1.609 \times 8.314}{42700} = \left(\frac{1}{T_2} - \frac{1}{281}\right)$$

$$-3.1 \times 10^{-4} + \frac{1}{281} - \frac{1}{T_2}$$

$$T_2 = 308.13 \text{ K}$$

$$T_2 = 35.12°\text{C}$$

Hence, the answer is (D).

Q7. One kmol of an ideal gas (C_p = 21 kJ/kmol.K) undergoes a constant pressure process from 300 K to 500 K. The molar entropy of the gas at 300 K is 150 kJ/kmol.K. The molar entropy (in kJ/kmol.K) at 500 K (up to 1 decimal place) is ________.

Solution:

$$\Delta s = C_p \ln\frac{T_2}{T_1} - R\ln\frac{P_2}{P_1}$$

$$s_{500} - s_{300} = C_p \ln\frac{T_2}{T_1}$$

$$s_{500} = 150 + 21\ln\left(\frac{500}{300}\right)\text{kJ/kmol.K} = 160.7 \text{ kJ/kmol.K}$$

Hence, the answer is 160.7

Q8. A spring, having a spring constant of 350 kN/m, is initially compressed by 0.4 cm. The work required (in J) to compress it by another 0.6 cm (up to 1 decimal place) is ________.

Solution:

Given:

x_1 = 0.4 cm

x_2 = (0.4 + 0.6) cm = 1 cm

k = 350 kN/m

W = Energy stored at final compression-Energy stored at initial compression

$$= \frac{kx_2^2}{2} - \frac{kx_1^2}{2}$$

$$=\frac{k}{2}\left(x_2^2-x_1^2\right)$$

$$=\frac{350}{2}\left\{\left(\frac{1}{100}\right)^2-\left(\frac{0.4}{100}\right)^2\right\}\times 1000\text{ J}$$

$$= 14.7\text{ J}$$

Hence, answer is 14.7

Q9. An ideal gas has a molar mass of 40 kg/kmol. Take $R = 8.314$ kJ/kmol.K. At a pressure of 2 bar and a temperature of 300 K, the volume (in m³) of 1 kg of this gas (up to 2 decimal places) is _______.

Solution:

$$\mathcal{M} = 40\frac{\text{kg}}{\text{kmol}}$$

$R = 8.314$

$p = 2\text{ bar} = 2 \times 10^5\text{ Pa}$

$T = 300\text{ K}$

$m = 1\text{ kg}$

Hence, the number of moles $n=\dfrac{m}{\mathcal{M}}=\dfrac{1}{40}\text{kmol}=\dfrac{1}{40}\times 1000\text{ mol}$

Now, the equation of state is

$$pV = nRT$$

$$V = \frac{nRT}{p}$$

$$=\frac{\frac{1}{40}\times 8.314\times 300\times 1000}{2\times 10^5}\text{m}^3$$

$$= 0.31\text{ m}^3$$

Hence, the answer is 0.31.

Q10. Consider the following statements for an ideal gas undergoing a reversible non-flow process:

P. If the process is adiabatic, the change in enthalpy of the gas is necessarily zero.

Q. If the process is adiabatic, the change in entropy of the gas is necessarily zero.

R. If the process is isothermal, the change in enthalpy of the gas is necessarily zero.

S. If the process is isothermal, the change in entropy of the gas is necessarily zero.

Which one of the following options is valid?

(A) Only P is correct (B) Only S is correct

(C) Only Q and R are correct (D) Only P and S are correct

Solution:

The enthalpy depends on the temperature and the process is adiabatic and so the process is isentropic.

Hence, the answer is (C).

Q11. An ideal Otto cycle (O) and an ideal diesel cycle (D) have the same maximum temperature and reject equal amount of heat. Also, the working fluid enters at the same state before compression. One of the following statements always true about their efficiencies (η) is

(A) $\eta_0 > \eta_D$ (B) $\eta_0 = \eta_D$

(C) $\eta_0 < \eta_D$ (D) $\eta_0 = 1 - \eta_D$

Solution:

For the same state before compression, an Otto cycle is more efficient than a Diesel cycle only at the same compression ratio. For the present case, the Diesel cycle has got higher compression ratio than the Otto cycle.

Hence, the answer is (C).

Q12. A reversible engine receives 75 kJ/s of energy from a reservoir at 750 K and does 12 kJ/s of work. The heat is rejected to two reservoirs at 650 K and 550 K. The rate of heat rejection (in kJ/s) to the reservoir at 650 K is

(A) 11 (B) 31

(C) 41 (D) 52

Solution:

Given:

The heat added at $T_1 = 750$ K, $Q_1 = 75$ kJ/s

Work done, $W = 12$ kJ/s

Let the heats rejected at 650 K ($=T_2$) and 550 K ($=T_3$) be Q_2 and Q_3, respectively.

Therefore, from the first law of Thermodynamics,

$$Q_2 + Q_3 = Q_1 - W$$

$$= (75 - 12)\ \text{kJ/s} \qquad \text{(i)}$$

Also for a reversible cycle,

$$\oint \frac{\delta Q}{T} = 0$$

$$\frac{Q_1}{T_1} - \frac{Q_2}{T_2} - \frac{Q_3}{T_3} = 0$$

$$\frac{75}{750} - \frac{Q_2}{650} - \frac{Q_3}{550} = 0$$

$$\frac{Q_2}{65} + \frac{Q_3}{55} = 1$$

$$55Q_2 + 65Q_3 = 3575 \qquad \text{(ii)}$$

Solving equations (i) and (ii) we get $Q_2 = 52$ kJ/s

Hence, the answer is (D).

Q13. A gas obeys the following equation of state:

$$P(v - b = RT + \frac{aP^2}{T})$$

where v is molar volume, and a, b are constants with values $a = 10^{-5}$ J.K/Pa^2.kmol and $b = 8 \times 10^{-2}$ m^3/kmol. Take $C_p = 30$ kJ/kmol.K. At 10 bar and 500 K, the value of the Joule-Thomson coefficient (in K/Pa) is

(A) -2×10^{-6} (B) -4×10^{-6} (C) 2×10^{-6} (D) 4×10^{-6}

Solution:

The Joule Thomson Coefficient is given by

$$\mu_j = \left(\frac{\partial T}{\partial p}\right)_h = \frac{1}{C_p}\left[T\left(\frac{\partial v}{\partial T}\right)_P - v\right]$$

$$= \frac{1}{C_p}\left[T\left(\frac{R}{P} - \frac{aP}{T^2}\right) - v\right]$$

Now,

$$v = b + \frac{RT}{p} + \frac{aP}{T} = 8\times10^{-2} + \frac{8.314\times500}{1000} + \frac{10^{-5}\times1000000}{500}\,\mathrm{m^3/kmol} = 4.257\ \mathrm{m^3/kmol}$$

$$\text{Therefore, } \mu_j = \frac{1}{C_p}\left[T\left(\frac{R}{P} - \frac{aP}{T^2}\right) - v\right]$$

$$= \frac{1}{30}\times\left[500\times\left(\frac{8.314}{1000} - \frac{10^{-5}\times100000}{500^2}\right) - 4.257\right]\mathrm{K/kPa}$$

$$= -4\times 10^{-6}\ \mathrm{K/Pa}$$

Hence, the answer is (B).

Q14. In an ideal Rankine cycle, steam enters the turbine at 10 MPa and 500 °C (h= 3375.1 kJ/kg, s = 6.5995 kJ/kg.K). It is cooled in the condenser at a pressure of 10 kPa. At 10 kPa, h_f = 191.81 kJ/kg, S_f = 0.6492 kJ/kg.K, h_g = 2583.9 kJ/kg and S_g = 8.1488 kJ/kg.K. The heat rejected in the condenser

(in kJ/kg) is

(A) 1898 (B) 3796

(C) 949 (D) 2847

Solution:

$S_1 = S_2$

$6.59 = S_f + x(S_g - S_f)$

$6.59 = 0.6492 + x(8.1488 - 0.6492)$

$x = 0.7921$

$h_f = h_3 = 191.81$

$h_2 = h_f + x(h_g - h_f)$

$= 191.81 + 0.7921(2583.9 - 191.81) = 2086.82$

Heat rejected through the condenser,

$h_2 - h_3 = 2086.82 - 191.81$

$= 1895.01$

≈ 1898

Hence, the answer is (A).

Q15. Methane has compressibility factor value of 0.9 at reduced pressure of 1.0 and reduced temperature of 1.5. For propane, the critical temperature and pressure are 369.8 K and 42.48 bar, respectively. Take $R = 8.314$ kJ/kmol.K. Applying the principle of corresponding states, the molar volume of propane (in m³/kmol) at the same reduced pressure and temperature is

(A) 0.355 (B) 0.526

(C) 0.791 (D) 0.977

Solution:

We know from the law of corresponding states that $Z_C = \frac{3}{8}$

Now $v_c = \frac{Z_c R T_c}{P_c} = \frac{3}{8} \times \frac{8.314 \times 369.8}{4248} \frac{\text{m}^3}{\text{kmol}} = 0.2714 \frac{\text{m}^3}{\text{kmol}}$

Also, $v_r = \frac{Z}{Z_c} \times \frac{T_r}{P_r} = \frac{0.9}{(3/8)} \times \frac{1.5}{1} = 3.6$

Therefore, $v = v_c \times v_r = 0.2714 \times 3.6 \frac{\text{m}^3}{\text{kmol}} = 0.977 \frac{\text{m}^3}{\text{kmol}}$

Hence, the answer is (D).

Q16. A rigid insulated vessel is divided into two compartments by a partition. One compartment contains 12 kg of oxygen at 200 kPa and 280 K. The other compartment contains 26 kg of carbon dioxide at 400 kPa and 360 K. The specific heats at constant volume in kJ/kg.K for oxygen and carbon dioxide are 0.662 and 0.653, respectively. The partition is removed and the gases are allowed to mix. Considering both gases are ideal, the final temperature (in K) of the mixture (up to 1 decimal place) is ________.

Solution:

$$T = \frac{m_1 C_1 T_1 + m_1 C_2 T_2}{m_1 C_1 + m_2 C_2}$$

$$= \frac{12 \times 0.662 \times 280 + 26 \times 0.653 \times 360}{12 \times 0.662 + 26 \times 0.653}$$

$$= \frac{2224.32 + 6112.08}{7.994 + 16.978}$$

$$= 334.499\text{K}$$

Hence, the answer is 334.5

Q17. Moist air having 60% relative humidity, enters a steady-flow air-conditioning unit at 102 kPa and 30 °C. The volume flow rate of the moist air entering the unit is 0.1 m³/s. The moist air leaves the unit at 95 kPa and 15 °C with a relative humidity of 100%. Liquid condensate leaves the unit at 15 °C.

For water: at 15 °C, h_f = 62.982 kJ/kg, h_g = 2528.3 kJ/kg, P_{sat} = 1.7057 kPa.
at 30 °C, h_f = 125.74 kJ/kg, h_g = 2555.6 kJ/kg, P_{sat} = 4.2469 kPa.

For air, specific heat at constant pressure is 1.004 kJ/kg.K and the specific gas constant is 0.287 kJ/kg.K.

Neglecting heat leakage to the surrounding, the magnitude of heat extracted (in kW) from the air stream (up to 2 decimal places) is ________.

Solution:

Let the partial pressure of water vapour at inlet be $(p_v)_1$.

$$\text{Hence, } \frac{(p_v)_1}{(p_{sat})_1} = \varphi_1$$

$$(p_v)_1 = 0.6 \times 4.2469 \text{ kPa}$$

$$= 2.548 \text{ kPa}$$

Hence, specific humidity at inlet, $\omega_1 = 0.622 \dfrac{(p_v)_1}{p_{atm} - (p_v)_1} = 0.0519$

Let the partial pressure of water vapour at inlet be $(p_v)_2$.

$$\text{Hence, } \frac{(p_v)_2}{(p_{sat})_2} = \varphi_2$$

$$(p_v)_2 = 1 \times 1.7057 \text{ kPa}$$

$$= 1.7057 \text{ kPa}$$

Hence, specific humidity at inlet, $\omega_2 = 0.622 \dfrac{(p_v)_2}{p_{atm} = (p_v)_2} = 0.0114$

The density of air at inlet, $\rho_1 = \dfrac{p_1}{RT} = \dfrac{102000}{287 \times 303} \text{kg/m}^3 = 1.184 \text{ kg/m}^3$

Therefore, the mass flow rate of air, $\dot{m} = \rho_1 Q_1 = 1.184 \times 0.1 \text{ kg/s} = 0.1184 \text{ kg/s}$

Now, the specific enthalpy of air at inlet, $h_1 = 1.005 t_1 + \omega_1 (2501 + 1.88 t_1)$

$= 70.81$ kJ/kg of d.a

Now, the specific enthalpy of air at inlet, $h_2 = 1.005\, t_2 + \omega_2\, (2501 + 1.88\, t_2)$

$= 43.9$ kJ/kg of d.a

Hence the rate of heat extracted from the air stream,

$$\dot{Q} = \dot{m}\left\{h_1 - h_2 - (\omega_1 - \omega_2)(h_f)_2\right\}$$

$$= 0.1184 \times \{70.81 - 43.9 - (0.0159 - 0.0114) \times 62.982\} \text{ kW}$$

$$= 3.06 \text{ kW}$$

Hence, the answer is 3.06 .

Q18. Air at 150 kPa and 323 K is filled in a rigid vessel of 0.05 m³ capacity. For air, assumed as an ideal gas, specific heat at constant volume is 0.7163 kJ/kg.K and the specific gas constant is 0.287 kJ/kg.K. Neglect kinetic and potential energy changes. If 30 kJ of heat is added, the final temperature (in K) of air (up to 1 decimal place) is __________.

Solution:

$$V = 0.05 \text{ m}^3$$

$$C_v = 0.7163 \text{ kg/kJK}$$

$$R = 0.287 \text{ kJ/kgK}$$

$$pV = mRT$$

$$m = \frac{150 \times 0.05}{0.287 \times 323} \text{ kg} = 0.0809 \text{ kg}$$

$$Q = 30 \text{ kJ added,}$$

$$Q = \Delta U + W$$

$30 = 0.0809 \times 0.7163 \times (T_2 - 323)$ (Since W=0 for this process)

$517.7 = T_2 - 323$

$T_2 = 840.7$ K

Hence, the answer is 840.7

Q19. Superheated steam at 2 bar and 300°C, with an enthalpy of 3072.1 kJ/kg, enters a horizontal adiabatic nozzle with negligible velocity and leaves at 0.2 bar as saturated vapour with an enthalpy of 2609.9 kJ/kg. Assuming steady flow and neglecting the potential energy changes, the exit velocity (in m/s) of the steam (up to 1 decimal place) is _________.

Solution:

Here,

$P_1 = 2$ bar

$T_1 = 300^\circ$C

$h_1 = 3072.1$ kJ/kg

$P_2 = 0.2$ bar

$h_2 = 2609.9$ kJ/kg

$$h_1 + \frac{v_1^2}{2} = h_2 + \frac{v_2^2}{2}$$

$$h_1 - h_2 = \frac{v_2^2}{2}$$

$$v_2 = \sqrt{2 \times (3072.1 - 2609.9) \times 1000}$$

$$v_2 = 961.457$$

Hence, the answer is 961.5.

Q20. A given mass of a simple compressible substance undergoes a reversible cycle, as shown in the *P-V* diagram. The magnitude of the net work done during the cycle is 3 kJ. The pressure (in bar) at point C (up to 1 decimal place) is ________.

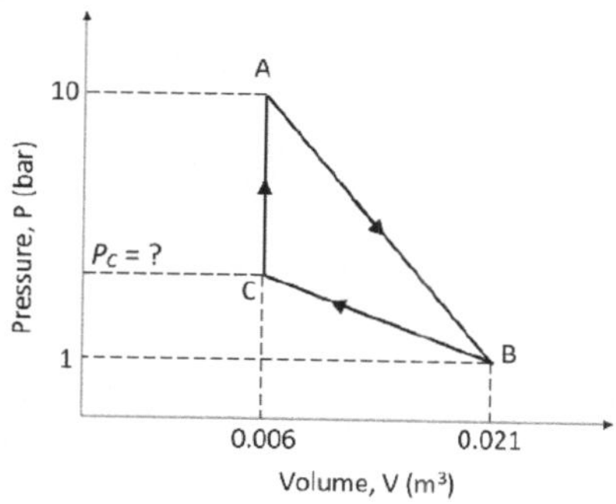

Solution:

Here the net work done is 3 kJ,

$$\left\{\frac{1}{2}(10-9)\times 0.015\times 100-\frac{1}{2}(P_c-1)\times 0.015\times 100\right\}=3$$

$$450-200=50\times(P_C-1)$$

$$P_c-1=\frac{250}{50}$$

$$P_C=5+1$$

$$P_C=6$$

Hence, answer is 6.

Q21. One kmol of an ideal gas at 300 K and 10 bar is reversibly heated in a constant volume process to 500 K. It is then reversibly and isothermally expanded to 2 bar. Take C_v = 20.8 kJ/kmol.K and R = 8.314 kJ/kmol.K. The total heat supplied (in kJ) to the gas (up to 1 decimal place) is ________.

Solution:

$n = 1$ kmol

$T_1 = 300$ K

$p_1 = 10 \text{ bar} = 10^6$ Pa

$T_2 = 500$ K

Constant volume heating,

$Q_1 = \Delta U$

$Q_1 = nC_v\Delta T$

$= 1 \times 20.8 \times (500 - 300)$ kJ

$= 4160$ kJ

Isothermal expansion,

$$p_1 V_1 = nRT_1$$

$$V_1 = V_2 = \frac{1\times 8.314\times 300\times 1000}{10^6}$$

$=2.492$

$$V_3 = \frac{1\times 8.314\times 500\times 1000}{2\times 10^6}$$

$= 20.785$

$$Q_2 = \Delta U + \int_{V_2}^{V_3} pdV$$

$$Q_2 = P_2V_2 \ln\frac{V_3}{V_2} \text{(Since } \Delta U = 0)$$

$$= 1\times 8.314\times 500\ln\frac{20.785}{2.492}$$

$= 8817.6$ kJ

So the total amount of heat supplied,

$Q = Q_1 + Q_2$

$= 4160+8817.6 = 12977.6$

Hence, the answer is 12977.6.

Q22. A rigid container is completely filled with a liquid having a constant isothermal compressibility of 1.09 x 10^{-4} bar^{-1} and a constant coefficient of volume expansion of 1.12 x 10^{-3} K^{-1}. The liquid is initially at 300 K and 1 bar. Heat is supplied to the liquid to raise its temperature to 350 K. Assuming that no phase change occurs, the final pressure (in bar) of the liquid (up to 1 decimal place) is _______.

Solution:

The fraction of compressed volume, $-\dfrac{\Delta V}{V} = \alpha_V \Delta t = 1.12\times10^{-3}\times50 = 0.056$

Now, the compressibility is defined as

$$K = -\frac{(\Delta V/V)}{\Delta p}$$

$$\Delta p = -\frac{\left(\dfrac{\Delta V}{V}\right)}{K} = \frac{0.056}{1.09\times10^{-4}}\text{bar} = 513.76 \text{ bar}$$

Hence the final pressure = (513.76 + 1) bar = 514.76 bar

Hence, the answer is 514.8.

GATE 2017 THERMODYNAMICS

Q1. Given $d\varphi = f(T)dT + (T/V)dV$ and $d\psi = Tdp + (T/p^2)\,dV$, then

(A) both φ and ψ *and* property

(B) neither φ nor ψ is a property

(C) φ is a property ψ but is not a property

(D) ψ is a property but φ is not a property

Solution:

$$d\varphi = f(T)\,dT + (T/V)dV \text{ and } d\psi = Tdp + (T/p^2)dV$$

For a given parameter to be a property, it should be an exact differential, i.e:, if z is described as a function of two variables x and y.

Then, $dz = \left(\frac{\partial z}{\partial x}\right)_y dx + \left(\frac{\partial z}{\partial y}\right)_x dy$

$dz = Mdx + Ndy$

Such that $\left(\frac{\partial M}{\partial y}\right)_x = \left(\frac{\partial N}{\partial x}\right)_y$

For $z = \varphi, M = f(T), N = \frac{T}{V}$

$$\therefore \left(\frac{\partial M}{\partial y}\right)_x = \left(\frac{\partial f(T)}{\partial V}\right)_T = 0$$

$$\left(\frac{\partial N}{\partial x}\right)_y = \left(\frac{\partial (T/V)}{\partial V}\right)_T = \frac{1}{V}$$

Hence, $\left(\frac{\partial M}{\partial y}\right)_x \neq \left(\frac{\partial N}{\partial x}\right)_y$

Therefore, φ is not a property.

Similarly, for ψ :

$$M = T, \quad N = \frac{T}{p^2}$$

$$\left(\frac{\partial T}{\partial V}\right)_p = 0, \quad \left(\frac{\partial (T/p^2)}{\partial p}\right)_V = -2 \times T \times P^{-3} = \frac{-2T}{p^3}$$

$\therefore \psi$ is also not a property.

This means neither φ, nor ψ is a property

Hence, the answer is (B).

Q2. A paddle wheel is installed in a rigid insulated tank containing 10 kg air (C_v= 0.718kJ/kg.K). A torque of 100 Nm is applied on the paddle wheel to rotate it at 60 revolution per minute for 2 minutes. At the end of the process, the increase in temperature of air in °C is

(A) 0 (B) 5.25

(C) 10.50 (D) 21.50

Solution:

m = 10 kg

C_v = 0.718 kJ/KgK

τ = 100 N.m

N = 60 rpm

T = 100 Nm

$\omega = 2 \times \pi \times N$

$= 2\pi \times 60$ rad/min = 377 rad/min

t = 2 minutes

Hence, the angle turned $\theta = \omega t = 377 \times 2$ rad = 754 rad

Hence the work done on the system, $W = T\theta = 100 \times 754$ J = 75400 J

From the first law of thermodynamics

$Q = \Delta U - W$ (For work done "on" the system)

$\Delta U - W = 0$ (Since rank is insulated, hence $Q = 0$)

$mC_v \Delta T = W$

$$\Delta T = \frac{W}{mC_v}$$

$$= \frac{75400}{10 \times 718} \text{°C}$$

$$= 10.5\text{°C}$$

Hence, the answer is (C).

Q3. Consider two systems each containing 20 kg of air at the same temperature and pressure. It is desired to increase the temperature of the air in both systems by 10°C. One system undergoes a constant pressure heat addition process and the other undergoes a constant volume heat addition. The difference in the values of heat transferred to the two systems in kJ is

(A) 30.5 (B) 44.2 (C) 57.5 (D) 73.2

Solution:

$m = 20$ kg

Temperature and pressure of both are the same same i.e. (T, p)

For the 1st case,

Constant pressure addition

$\delta Q_1 = dH\text{-}vdp$ $(vdp = 0)$

$\delta Q_1 = dH = mC_p dT$

$Q_1 = \Delta H = mC_p \Delta T$

Constant-volume addition

$\delta Q_2 = dU + dW,\ dV = 0\ (dW = pdV)$

$\delta Q_2 = mCvdT$

$Q_2 = \Delta U = mCv\Delta T$

$\therefore Q_1\text{-}Q_2 = \text{m}(Cp\text{-}Cv)\ \Delta T$

$= 20 \times 10 \times R$

For air, $R = 0.287$ kJ/kg.K

$= 20 \times 10 \times 0.287 = 57.4$ kJ/kg.K

Hence, the answer is (C).

Q4. A refrigerator is used to maintain certain space at 10°C. It pumps 18000 kJ/hour of heat from the space to the atmosphere at 30°C. If the power input to the refrigerator is 2 kW, the ratio of COP of this refrigerator to that of a Carnot refrigerator (up to 2 decimal places) is ______.

Solution:

For refrigerator,

$$\text{COP} = \frac{Q_2}{W_1} = \frac{\text{amount of heat removed}}{\text{work input}}$$

$$= \frac{18 \times 10^3 \text{ kJ/hr}}{2 \text{ kJ/s}} = \frac{18 \times 10^3 \text{ kJ/s}}{3600 \times 2 \text{ kJ/s}} = 2.5$$

For Carnot refrigerator,

$$\text{COP} = \frac{T_2}{T_1 - T_2} = \frac{283}{303 - 283} = 14.15$$

$$\frac{(\text{COP})_{ref}}{(\text{COP})_{Carnot\ ref}} = \frac{2.5}{14.15} = 0.176$$

Hence, the answer is 0.176.

Q5. A thermal cycle receives 2000 kJ of heat from a heat source at 1000 K. It rejects 300 kJ of heat to a heat sink at 300 K and also rejects 250 kJ of heat to another heat sink at 200 K during the cycle. The cycle is

(A) reversible (B) irreversible (C) impossible (D) work absorbing

Solution:

For a reversible cycle,

$$\oint \frac{dQ}{T} = 0$$

For irreversible cycle,

$$\oint \frac{dQ}{T} < 0$$

Now for the present cycle,

$$\oint \frac{dQ}{T} = \frac{Q_1}{T_1} + \frac{Q_2}{T_2} + \frac{Q_3}{T_3}$$ (Q_2 and Q_3 are heat rejected to the sink, hence they are -ve)

$$= \frac{2000}{1000} + \left(\frac{-250}{200}\right) + \left(\frac{-300}{300}\right)$$

$$= 2 - 1.25 - 1$$

$$= 2 - 2.25$$

$$= -0.25 < 0$$

∴ The process is irreversible

Hence, the answer is (B).

Q6. Saturated liquid water is slowly heated at a constant pressure of 200 kPa to a final state where its quality reaches 0.65. For water at 200 kPa: T_{sat} = 120.23°C, h_f = 504.68 kJ/kg, h_g = 2706.60 kJ/kg. The change in the specific entropy in kJ/kgK is

(A) 3.04 (B) 3.24 (C) 3.44 (D) 3.64

Solution:

Initial state, water is saturated liquid, i.e. the initial dryness fraction $x_1 = 0$

$p_1 = 200$ kPa; $T_{sat} = 120.23$°C, $h_f = 504.68$ kJ/kg,

$h_g = 2706.60$ kJ/kg

The final dryness fraction, $x_2 = 0.65$

At the initial state, $h_1 = h_f = 504.68$ kJ/kg

The specific entropy at state 2 is given by,

$$s_2 = s_f + \frac{x_2\left(h_g - h_f\right)}{T_{sat}}$$

Hence, the change in specific entropy

$$\Delta s = s_2 - s_f$$

$$= \frac{x_2(h_g - h_f)}{T_{sat}}$$

$$= \frac{0.65 \times (2706.6 - 504.68)}{120.23 + 273} \text{ kJ/kg. K}$$

$$= 3.64 \text{ kJ/kg.K}$$

Hence, the answer is (D).

Q7. Given the thermodynamic functional relations: $p = p(v,T)$ and $T = T(p,v)$, the term $\left.\frac{\partial p}{\partial v}\right|_T \times \left.\frac{\partial v}{\partial T}\right|_p$ is

(A) $-\left(\left.\frac{\partial T}{\partial p}\right|_v\right)^{-1}$ (B) $\left(\left.\frac{\partial T}{\partial p}\right|_v\right)^{-1}$ (C) $\left(\left.\frac{\partial T}{\partial p}\right|_v\right)$ (D) 1

Solution:

$p = p(v,T)$, $T = T(p,v)$

∴ a relation exits between the 3 variables

$$\therefore \left(\frac{\partial p}{\partial v}\right)_T \times \left(\frac{\partial v}{\partial T}\right)_p \times \left(\frac{\partial T}{\partial P}\right)_v = -1$$

$$\therefore \left(\frac{\partial p}{\partial v}\right)_T \times \left(\frac{\partial v}{\partial T}\right)_p = -1 \times \left\{\left(\frac{\partial T}{\partial p}\right)_v\right\}^{-1} = -\left\{\left(\frac{\partial T}{\partial p}\right)_v\right\}^{-1}$$

Hence, the answer is (A).

Q8. Two closed cycle gas turbine engines, A and B, operate on air standard Brayton cycle with efficiencies of η_A and η_B, respectively. If they operate between the same maximum temperatures, but with different pressure ratios of r_{pA} and r_{pB}, ($r_{pA} > r_{pB}$), then,

(A) $\eta_A = \eta_B$

(B) $\eta_A > \eta_B$

(C) $\eta_A < \eta_B$

(D) Cannot be determined as the efficiencies only at the optimal r_p values.

Solution:

$$\eta_{\text{Brayton}} = 1 - \frac{1}{\left(r_p\right)^{(\gamma-1)/\gamma}}$$

$$\eta \propto r_{p,} \text{ i.e., } \eta \uparrow, \uparrow$$

$$r_{pA} > r_{pB}$$

$$\therefore \eta_A > \eta_B$$

Hence, the answer is (B).

Q9. The values of density and isentropic compressibility of water at certain pressure and temperature are given as 1000 kg/m³ and 40×10⁻¹⁰ Pa⁻¹, respectively. The speed at which sound travels in water under these conditions in m/s is equal to________.

Solution:

$\rho = 1000$ kg/m^3

$k = 40\times10^{-10}$ Pa^{-1}

Speed of sound $= \sqrt{\dfrac{1}{\chi\rho}} = \sqrt{\dfrac{1}{40 \times 10^{-10} \times 10^3}} = \sqrt{\dfrac{10^{10}}{4 \times 10^4}} = 500$ m/s

Hence, the answer is 500.

Q10. Length of a certain metal rod at 0°C is 10 cm. The coefficient of linear expansion of that metal varies with temperatures as $10^{-4} + 10^{-5} \times T$ (cm/cm)/°C. When the length of the metal rod is 10.2 cm, the rise in temperature in °C is________.

Solution:

$L = L_0(1 + \alpha\Delta T)$

$L - L_0 = L_0\,\alpha\Delta T$

In the differential form,

$dL = L_0\,\alpha dT$

$dL = L_0 \times (10^{-4} + 10^{-5} \times T) \times dT$

$$\int_{L=10}^{L=10.2} dL = \int_{T=0}^{T} L_0 \times \left(10^{-4} + 10^{-5} \times T\right) \times dT$$

$$10.2 - 10 = 10 \times \left[10^{-4} \times T + 10^{-5} \times \frac{T^2}{2}\right]_{T=0}^{T}$$

$$\frac{T^2}{200000} + \frac{T}{10000} = 0.02$$

$$T^2 + 20T - 4000 = 0$$

Solving the quadratic equation we get, $T = 54.03°C$

Therefore, the rise in temperature is $\Delta T = (54.03\text{-}0)°C = 54.03°C$

Hence, the answer is 54.03.

Q11. In a polytropic compression process, one kg of an ideal gas having a molecular weight of 40 kg/kmol is compressed from 100 kPa, 300 K to 400 kPa, 360 K. The magnitude of the work in kJ for the process is

(A) 52.3 (B) 62.3 (C) 72.3 (D) 82.3

Solution:

$m = 1$ kg, $M = 40$ kg/kmol

$p_1 = 100$ kpa, $T_1 = 300$K

$p_2 = 400$ kpa, $T_2 = 360$k

$p_2V_2 = nRT_2$

$p_1V_1 = nRT_1$

$100 \times V_1 = \frac{1}{40} \times 8.314 \times 300$

$V_1 = 0.62355 \text{ m}^3$

$$V_2 = \frac{\frac{1}{40} \times 8.314 \times 360}{400}$$

$V_2 = 0.187 \text{ m}^3$

For polytrophic compression process

$$W = \frac{p_1V_1 - p_2V_2}{r - 1}$$

Now,

$$\frac{T_2}{T_1} = \left(\frac{p_2}{p_1}\right)^{r-1/r},$$

$$\frac{360}{300} = \left(\frac{400}{100}\right)^{r-1/r}$$

$$\ln 1.2 = \frac{r-1}{r} \ln 4$$

$$\frac{\ln 1.2}{\ln} = \frac{r-1}{r}$$

$$0.131 = \frac{r-1}{r}$$

$$r = 1.15$$

$$\therefore W = \frac{100 \times 0623 - 400 \times 0.187}{1.15 - 1}$$

$$= 82.3\text{kJ}$$

Hence, the answer is (D).

Q12. Two streams of air (C_p=1005 J/Kg.K) flow through insulated pipes 1 and 2 with the conditions as shown in figure. They mix in an insulated pipe-3 and the mixture steadily exits with a velocity of 100 m/s at 150 kPa. Neglecting the change in potential energy in all the pipes, the exit area of the pipe-3 in m² (upto 3 decimal places) is_________.

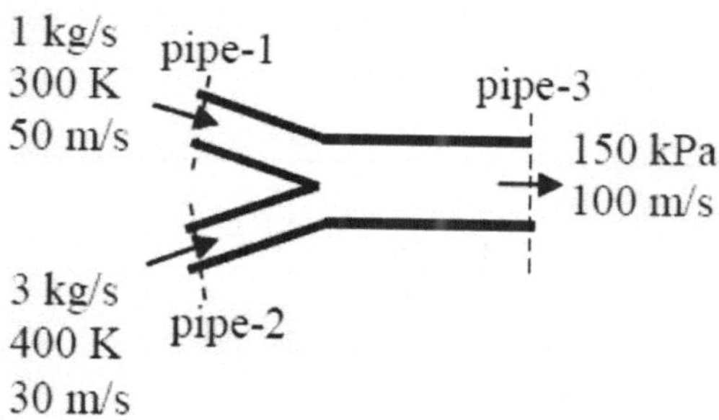

Solution:

$$\dot{m} = \frac{C \times A}{v}$$

$\dot{m}$ = mass flow rate in (kg/s)

C = velocity in m/s

A = area in m²

v = specific volume (m³/kg)

$$\dot{m}_3 = \dot{m}_1 + \dot{m}_2 = 3 + 1 = 4 \text{ kg/s}$$

Since system is insulated

$$\therefore \dot{m}_1 C_p T_1 + \dot{m}_2 C_p T_2 = \dot{m}_3 C_p T_3$$

$$\dot{m}_1 T_1 + \dot{m}_2 T_2 = \dot{m}_3 T_3$$

$$\Rightarrow 300 + 1200 = 4 \times T_3$$

$$T_3 = 375\ K$$

$$p_3 V_3 = n_3 R T_3$$

$$150 \times V_3 = \frac{m_3}{29} \times 8.314 \times 375$$

$$v_3 = \frac{v_3}{m_3} = \frac{0.287 \times 375}{150} \text{ m}^3/\text{kg}$$

$v_3 = 0.717$ m³/kg

$$\therefore \text{Area} = \frac{0.717 \times 4}{100} = 0.0287 \text{ m}^2$$

Hence, the answer is 0.029.

Q13. A 1 m³ rigid vessel contains air at 200 kPa. A vacuum pump is connected to the vessel in order to control the pressure inside. The volume flow rate of air through the pump is maintained at a constant value of 0.1 m³/s. If the pump operates for 10 seconds and the temperature of the air is maintained constant during operation, the pressure in the tank in kPa after 10 seconds (up to 2 decimal places) is________.

Solution:

$$Q = \frac{\dot{m}}{\rho}$$

$$-\dot{m} = Q\rho$$

$$-\frac{dm}{dt} = Q\frac{m}{V}$$

$$-\frac{dm}{m} = \frac{Q}{V}dt$$

$$-\int_{m_1}^{m_2} \frac{dm}{m} = \int_0^{10} \frac{0.1}{1} dt$$

$$\ln\left(\frac{m_2}{m_1}\right) = -0.1 \times 10$$

$$\frac{m_2}{m_1} = 0.3678$$

Now, $pV = mRT$. Hence $\frac{p_2}{p_1} = \frac{m_2}{m_1} \times \frac{V_1}{V_2} \times \frac{T_2}{T_1}$.

For the tank V and T are constant. Therefore,

$$\frac{p_2}{p_1} = \frac{m_2}{m_1} = 0.3678$$

$p_2 = 0.3678 \times 200 \text{ kPa} = 73.56 \text{ kPa}$

Hence, the answer is 73.56.

Q14. A heat engine receives Q_1 kJ of heat from a hot reservoir and rejects Q_2 kJ of heat to a cold reservoir. The work delivered by the heat engine is entirely supplied to a heat pump, which receives Q_3 kJ of heat from another reservoir and rejects Q_4 kJ of heat to the same cold reservoir. If the efficiency of the heat engine is 0.4 and COP of the heat pump is 4.0 the value of $(Q_2 + Q_4)/Q_1$ (up to 1 decimal place) is________.

Solution:

For heat engine,

$$\eta = \frac{W}{Q_1} \Rightarrow \frac{Q_1 - Q_2}{Q_1} = 0.4$$

$$\therefore 1 - \frac{Q_2}{Q_1} = 0.4$$

$$\therefore Q_2 = 0.6Q_1$$

$$\frac{Q_2}{Q_1} = 0.6$$

$$Q_1\text{-}0.6Q_1 = Q_4 - 0.75Q_4$$

$$\text{COP h.p} \Rightarrow \frac{Q_4}{Q_4 - Q_3} = 4$$

$$\therefore \frac{Q_4 - Q_3}{Q_4} = 0.25$$

$$1 - Q_3/Q_4 = 0.25 \quad \therefore Q_3/Q_4 = 0.75$$

$$Q_1 - 0.6Q_1 = Q_4 - 0.75Q_4$$

$$0.4Q_1 = 0.25Q_4$$

$$\therefore \frac{Q_4}{Q_1} = 1.6$$

$$\therefore\left(\frac{Q_2}{Q_1}+\frac{Q_4}{Q_1}\right)=1.6+0.6=2.2$$

Hence, the answer is 2.2.

Q15. A block of ice of mass 2 kg at 0°C is dropped into an insulated vessel containing 10 kg of liquid water at 25°C. The latent heat of melting of ice is 330 kJ/kg and specific heat of water is 4.2 kJ/kg.K. The change in the entropy of the universe in kJ/K (up to 3 decimal places) is________.

Solution:

Given:

$m_i = 2$ kg

$m_w = 10$ kg

$$L_i = 330 \text{ kJ/kg}$$

$$C_w = 4.2 \text{ kJ/kg.K}$$

$$(t_{init})_i = 0°C$$

$$(t_{init})_w = 25°C$$

$$(T_{init})_i = 273 \text{ K}$$

$$(T_{init})_w = 298 \text{ K}$$

Let the final temperature after mixing be t_f.

Now, heat gained by ice=Heat lost by water

$$m_i L_i + m_i C_w \{t-(t_{init})_i\} = m_w C_w \{(t_{init})_w - t_f\}$$

$$2 \times 330 + 2 \times 4.2 \times \{t_f - 0\} = 10 \times 4.2 \times (25 - t_f)$$

$$660 + 8.4\, t_f = 1050 - 42\, t_f$$

$$t_f = 7.74°C$$

$$T_f = 280.74 \text{ K}$$

Hence the change is entropy of the universe,

$$\Delta_S = (\Delta_S)_i + (\Delta S)_w$$

$$= \frac{m_i L_i}{(T_{init})_i} + \int_{(T_{init})_w}^{T_f} \frac{m_i C_w dt}{T} + \int_{(T_{init})_w}^{T_f} \frac{m_w C_w dT}{T}$$

$$= 0.5 \times 0.897 \times 1000 \ln\left(\frac{504.28}{353}\right) \text{ J/K}$$

$$= 0.147 \text{ kJ/K}$$

Hence, the answer is 0.147.

Q16. A pure substance (C_v = 0.733 kJ/kg.K) undergoes a reversible process in which its temperature increases linearly from 40°C to 85°C and its specific entropy increases by 600 J/kg.K. The work done by the system in kJ/kg is

(A) 160.2 (B) 164.3

(C) 168.3 (D) 172.3

Solution:

C_v = 0.733 kJ/kg.k

T_1 = 40°C

T_2 = 85°C

ΔS = 600 J/kg.K

$$ds = \frac{dq}{T} = \frac{CpdT}{T}$$

$$\Delta s = Cp \ln \frac{T_2}{T_1}$$

$$600 = Cp \times \ln\left(\frac{85+273}{40+273}\right)$$

C_p = 4.47 kJ/kg.K

$\delta q = du + \delta w$

$C_p \Delta T = C_v \Delta T + \delta w$

$\delta w = (C_p - C_v) \times \Delta T$

= (4.47-0.733) × 45kJ/Kg=168.165kJ/kg

Work done = 168.165 kJ/kg

Hence, the answer is (C).

Q17. An ideal gas having a mass of 0.5 kg is initially at 300 kPa, 80°C and occupies a volume of 0.14 m³. The gas undergoes an adiabatic process, where 50 kJ work is transferred to the gas. The pressure and volume at the final state are 300 kPa and 0.20 m³. The change in the entropy of the gas in J/K is

(A) 160.3 (B) 175.3 (C) 190.3 (D) 195.3

Solution:

$\dot{m}$ = 0.5 kg

p_1 = 300 kPa, T = 80°C = 353 K, V_1 = 0.14 m³

W = -50 kJ

p_2 = 300 kPa, V_2 = 0.20 m³

$$\frac{p_1V_1}{T_1} = \frac{p_2V_2}{T_2}$$

$Q = 0$ $$\frac{300 \times 0.14}{353} = \frac{300 \times 0.2}{T_2}$$

$\Delta U = -W$ $T_2 = 504.28$ K

$\Delta U = 50$

$50 = mC_v \Delta T$

$50 = 0.5 \times C_v \times (504.28\text{-}353)$

$C_v = 0.661$ kJ/kg.K

$C_p\text{-}C_v = R$

$pV = mRT$

$\therefore 300 \times 0.14 = 0.5 \times R \times 353$

$R = 0.237$ kJ/kg.K

$\therefore C_p = C_v + R = (0.237+0.661)$ kJ/kg.K $= 0.898$ kJ/kg.K

$$\Delta s = mCp \ln \frac{T_2}{T_1}$$

$$= 0.5 \times 0.897 \times 1000 \ln\left(\frac{504.28}{353}\right) \text{ J/K}$$

$$= 160.14 \text{ J/K}$$

Hence, the answer is (A).

Q18. The van der Waals equation of state is given as,

$\left(p + \frac{a}{v^2}\right)(v - b) = R_u T$, where p in bar, v in m^3/kmol and T is in K.

For air, the constants, a and b are 1.368 bar (m^3/kmol)^2 and 0.0367 m^3/kmol, respectively. Air is contained in a system at 160 K and 0.08 m^3 / kmol. If p_1 is the pressure calculated using ideal gas equation of state and p_2 is pressure calculated using van der Waals equation of state, then p_1/p_2 is equal to

(A) 1.78 (B) 1.52 (C) 1.28 (D) 1.0

Solution:

$$\left(p + \frac{a}{v^2}\right)(v - b) = R_u T$$

1 bar = 10^5 =10^2 kPa

$$\therefore \left\{p_{vw} + \frac{1.368 \times 10^2}{(0.08)^2}\right\} \times (0.08 - 0.0367) = 8.314 \times 160$$

$\therefore p_{vW} = 9346.47$ kPa

$$P_{ideal} = \frac{RuT}{v}$$

$$\text{For ideal gas, } p_{ideal} = \frac{8.314 \times 160}{0.08} \text{kPa} = 16628 \text{ kPa}$$

$$\therefore \frac{p_{ideal}}{p_{vw}} = \frac{16628}{9346.47} = 1.78$$

Hence, the answer is (A).

Q19. The values of specific volume of H_2O at 100ºC for saturated vapour states are 0.001044 m³/kg and 1.673 m³/kg, respectively. The slope of saturation pressure versus temperature curve, i.e.,$(dp / dT)_{sat}$ is 3570 Pa/K. The change in enthalpy in kJ/kg between the two saturation states is_________.

Solution:

$v_f = 0.001044$ m³/kg

$v_g = 1.673$ m³/kg

$$\left(\frac{dp}{dT}\right)_{sat} = 3570 \text{ Pa/K}$$

$$h_2 - h_1 = \frac{3570 \times (1.673 - 0.001044) \times 373}{1000}$$

$$= 2225.87 \text{ kJ/kg}$$

Hence, the answer is 2225.87

Q20. In a steam power plant, steam is first expanded isentropically in a turbine from an initial condition of 100 bar and 500°C to a pressure of 40 bar. Then the steam is reheated up to 500°C at constant pressure. The steam is then expanded isentropically in another turbine up to a condenser pressure 0.01 bar. For steam, at 100 bar, 500°C; h = 3373.7 kJ/kg, s = 6.5966 kJ/kg.K; at 40 bar, 500°C; h = 3445.3 kJ/kg, s = 7.0901 kJ/kg.K and at 0.01 bar; h_f = 29.3 kJ/kg, h_g = 2514.2 kJ/kg, s_f = 0.1059 kJ/kg.K, s_g = 8.9756 kJ/kg.K. The dryness fraction at the condenser inlet (up to 2 decimal places) is __________.

Solution:

$p_1 = 0.01$ bar

$h_f = 29.3$ kJ

$h_g = 2514.2$ kJ/kg

$s_f = 0.1059$ kJ/kg.K

$s_g = 8.9756$ kJ/kg.K

$s_1 = s_2$

$s = 7.0901$ kJ/kg k

Steam is wet at condenser inlet

$7.0901 = xs_g + (1\text{-}x)s_f$

$7.0901 = 8.9756x + (1\text{-}x) \times 0.1059$

$x = 0.78$

Hence, the answer is 0.78.

Q21. Air contains by volume 79% N_2 (molecular weight = 28 kg/kmol) and 21% O_2 (molecular weight = 32 kg/kmol). A stream of air flows at 32°C, 1 bar, at a rate of 2 m³/s and is mixed with another stream of O_2 flowing at 0.4 kg/s. The molecular weight of the mixture (up to 2 decimal places) is __________.

Solution:

We know that 1 kmol =22400 m³

Hence, molecular weight of N_2 = 28/22400 kg/m³ = 0.00125 kg/m³

And molecular weight of O_2=32/22400 kg/m³ = 0.00143 kg/m³

Hence the mass of N_2 present in 1 m³ of air, $(m)_{N_2}|_V = 0.79 \times 0.00125$ kg $= 9.875 \times 10^{-4}$ kg

And the mass of O_2 present in 1 m³ of air, $(m)_{O_2})|_V = 0.21 \times 0.00143$ kg $= 3.003 \times 10^{-4}$ kg

Therefore, the mass of N_2 present in 1 kg of air,

$$(m)_{N_2}\Big|_m = \frac{9.875\times10^{10-4}}{(9.875+3.003)\times10^{-4}}\ \text{kg} = 0.7668\ \text{kg}$$

And the mass of O_2 present in 1 kg of air,

$$(m)_{O_2}\Big|_m = \frac{3.003\times10^{-4}}{(9.875+3.003)\times10^{-4}}\ \text{kg} = 0.2332\ \text{kg}$$

The number of moles of N_2 present in 1 kg of air,

$$(n)_{N_2} = \frac{0.7668}{28}\ \text{kmol} = 0.0274\ \text{kmol}$$

The number of moles of O_2 present in 1 kg of air,

$$(n)_{O_2} = \frac{0.2332}{32}\ \text{kmol} = 0.0073\ \text{kmol}$$

Therefore, the mole fraction of N_2, $(X)_{N2} = \dfrac{0.0274}{0.0274+0.0073} = 0.7896$

And the mole fraction of $O_2, (X)_{O_2} = \dfrac{0.0073}{0.0274 + 0.0073} = 0.2104$

Hence, the molecular weight of air is given by

$$\mathcal{M}_{air} = (X)_{O_2}\,(\mathcal{M})_{O_2} + (X)_{N_2}\,(\mathcal{M})_{N_2}$$
$$= 0.2104 \times 32 + 0.7896 \times 28 \text{ kg/kmol}$$
$$= 28.84 \text{ kg/kmol}$$

Hence, the characteristic gas constant of air is

$$R_{air} = \frac{\overline{R}}{\mathcal{M}_{air}} = \frac{8314}{28.84} \text{J/kmol. K} = 280.28 \text{ J/kmol. K}$$

Now, let the mass flow rate of air at 32^{0}C (= 305 K), 1 bar, at a flow rate of Q = 2 m^3/s be $\cdot_{air}$. Therefore,

$$pQ = (\cdot_{air})\, R_{air}\, \mathrm{T}$$

$$\dot{m}_{air} = \frac{pQ}{R_{air}T} = \frac{10^5 \times 2}{288.28 \times 305}\frac{\text{kg}}{\text{s}} = 2.275 \text{ kg/s}$$

This stream mixes with another stream of O_2 flowing at 0.4 kg/s.

The number of moles of air flowing per second in the mixture,

$$(n)_{air} = \frac{2.275}{28.84} \text{ kmol} = 0.0789 \text{ kmol}$$

The number of moles of O_2 flowing per second in the mixture,

$$(n)_{air} = \frac{0.4}{32} \text{ kmol} = 0.0125 \text{ kmol}$$

Therefore, the mole fraction of air in the mixture, $(X)_{air} = \dfrac{0.0789}{0.0789 + 0.0125} = 0.8632$

And the mole fraction of $O_2, (X)_{O_2} = \dfrac{0.0125}{0.0789 + 0.0125} = 0.1368$

Hence, the molecular weight of the mixture

$$\mathcal{M}_{mixture} = (X)_{air}\,(M)_{air} + (X)_{O_2}\,(M)_{O_2}$$
$$= 0.8632 \times 28.84 + 0.1368 \times 32 \text{ kg/kmol}$$
$$= 29.27 \text{ kg/kmol}$$

Hence, the answer is 29.27.

Q22. Moist air enters a duct at a rate of 3 kg/s at 10°C, 80% relative humidity. The air is heated as it flows through the duct and exits at 30°C. No moisture is added or removed and the pressure of air in the duct is constant at 1 bar. The saturation vapour pressure (p_g) of H_2O at 10°C is 0.01228 bar. Specific enthalpy values of dry air at inlet and outlet of the duct are respectively 283.1 kJ/kg and 303.2 kJ/kg. The corresponding specific enthalpy values for water vapour are 2519.8 kJ/kg and 2556.3 kJ/kg. For steady state operation the amount of heat added to the moist air in kW (up to 2 decimal places) is __________.

Solution:

$\dot{m} = 3 kg/s$

$T_1 = 10°C$

RH = 80%

$T_2 = 30°C$

$(p_v)_{sat}$ 10°C is 0.01228 bar

Now, the enthalpy of moist air is given by,

$h = h_{da} + wh_g$

$$w = 0.622 \frac{p_v}{p_t - p_v}$$

$$\frac{80}{100} = \frac{pv}{(p_v)_{sat}} = \frac{p_v}{0.01228}$$

$$w = 0.622 \times \frac{0.0009824}{1 - 0.0009824} = 0.0061$$

$p_v = 0.009824$ bar

Therefore, the specific enthalpies of air at inlet and outlet are

$h_{in} = 283.1 + 0.0061 \times 2519.8$ kJ/kg= 298.47 kj/kg

$h_{out} = 303.2 + 0.0061 \times 2556.3$ kJ/kg = 318.79 kJ/kg

Hence, the amount of heat added is

$Q = \dot{} (h_{out} - h_{in}) = 3 \times (318.79 - 298.47)$kW = 60.97 kW

Hence, the answer is 60.97 .

GATE 2016
THERMODYNAMICS

Q. 1 – Q. 9 carry one mark each.

Q1. Which of the following thermodynamic properties is NOT an intensive property of a thermodynamic system:

(A) Pressure
(B) Temperature
(C) Density
(D) Volume

Solution:

Volume depends on the extent or mass of the system.

Hence, the answer is (D).

Q2. An U-tube manometer shows a height difference of z_1 between the two columns for a known gauge pressure P_1 (both z_1 and P_1 in appropriate units). If the height difference between the two columns is $2z_1$, then the corresponding gauge pressure will be:

(A) $P_1/2$
(B) $2P_1$
(C) P_1
(D) $4P_1$

Solution:

Height difference = z_1

Gauge pressure = P_1

∴ The gauge pressure for a height difference = $2z_1$

$P_2 = \rho g(2z_1) = 2\rho g z_1 = 2P_1)$

$P_2 = 2P_1$

Hence, the answer is (B).

Q3. Water vapour can be treated as an ideal gas,

(A) for all temperature and pressure

(B) for sufficiently low pressure, regardless of its temperature

(C) for very high pressure only

(D) for sufficiently low temperature, regardless of its pressure

Solution:

For sufficiently low pressure, regardless of its temp.

Real gases behave like ideal gas at sufficiently low pressure.

Hence, the answer is (B).

Q4. The thermal efficiency of a Carnot engine is 0.5. If the temperature of the cold reservoir is 300K, then the temperature of the hot reservoir is:

(A) 600 K (B) 1200 K

(C) 900 K (D) 450 K

Solution:

$\eta = 0.5$

$T_L = 300\text{K}$

$$\eta = 1 - \frac{T_L}{T_H}$$

$$0.5 = 1 - \frac{300}{T_H}$$

$$\frac{300}{T_H} = 0.5$$

$$\therefore T_H = \frac{300}{0.5} = 600 \text{ K}$$

Hence, the answer is (A).

Q5. In a reversible, constant-pressure, non-flow process, heat input is given by

(A) change in internal energy

(B) change in enthalpy

(C) change in entropy

(D) work output

Solution:

In reversible, constant- pressure non flow process

$Q = \Delta U + W$

$$Q = \Delta U + \int_1^2 pdV$$

$$Q = \Delta U + p\int_1^2 dV$$

$= U_2 - U_1 + p\,(V_2 - V_1)$

$= (U_2 + pV_2) - (U_1 + pV_1)$

$= H_2 - H_1$

$Q = \Delta H$

Hence, the answer is (B).

Q6. Moist air undergoes an adiabatic saturation process such that the relative humidity of air increases. For this process,

(A) Dry bulb temperature increases, specific humidity increases

(B) Dry bulb temperature increases, specific humidity decreases

(C) Dry bulb temperature decreases, specific humidity increases

(D) Dry bulb temperature decreases, specific humidity decreases

Solution:

In adiabatic saturation process, sensible cooling + Humidification occurs.

∴ Dry bulb temperature decreases and absolute humidity increases.

Hence, the answer is (C).

Q7. A steadily flowing ideal gas undergoes adiabatic throttling, where

T_1: temperature before throttling

T_2: temperature after throttling

Assuming no change in kinetic and potential energy due to throttling, which of the following is correct:

(A) $T_1 = T_2$ (B) $T_1 > T_2$

(C) $T_1 < T_2$ (D) $T_1 = \gamma T_2$, γ = Specific heat ratio

Solution:

In throttling, $H_1 = H_2$

For ideal gas

$H = mc_p dT$

∴ Enthalpy = constant, then T = constant

$T_1 = T_2$

Hence, the answer is (A).

Q8. For irreversible heat transfer from a hot body to a cold body, if Δ denotes the property change of both hot and cold bodies (i.e., difference between its final and initial values), then

(A) ΔS=0 (B) ΔU>0 (C) ΔS<0 (D) ΔS>0

Solution:

For irreversible heat transfer, $\Delta S > 0$

Hence, the answer is (D).

Q9. A closed system undergoes a cyclic process. For the net work done by the system on the surroundings, which of the following statements is FALSE:

(A) Net work is always zero

(B) Net work is $\oint PdV$ if the process is reversible

(C) Net work can be negative

(D) Net work can be positive

Solution:

Net work done by the system is not necessarily zero.

Hence, the answer is (A).

Q. 10 – Q. 22 carry two marks each.

Q10. Consider the following statements related to the second law of thermodynamics:

P. A cyclic heat engine cannot produce net work by exchanging heat only with one reservoir.

Q. The efficiency of a reversible heat engine is dependent on the nature and amount of working substance undergoing the cycle.

R. It is impossible to have a cyclic device which will produce no effect other than the transfer of heat from a cold body to a hot body.

S. It is impossible to have heat engines operating between a heat source and sink to have a lower efficiency than that of a reversible heat engine operating between the same source and sink.

For which of the following options, BOTH the statements are inconsistent with the second law of thermodynamics:

(A) P and R (B) P and Q

(C) R and S (D) Q and S

Solution:

Q. According to corollary of Carnot's theorem, the η of a reversible is independent of the nature & amount of working substance undergoing the cycle.

S. If heat engine are operating between the same source and sink, η of a rev HE is the maximum. (Carnot's theorem)

Hence, the answer is (D).

Q11. Consider the following statements related to air-standard Otto, Diesel, and Brayton cycles:

P. Brayton cycle has at least one isentropic and one isobaric process.

Q. Otto cycle has at least one isentropic and one isochoric process.

R. Diesel cycle has at least one isentropic and one isothermal process.

S. At least one of the cycles has an isothermal process.

For which of the following options, BOTH the statements are consistent with the operation of the above cycles:

(A) P and R (B) P and Q (C) R and S (D) P and S

Solution:

Brayton cycle

Two isentropic process

Two isobaric process

Otto Cycle

2 isochoric process

2 isentropic process

Hence, the answer is (B).

Q12. Volumetric analysis of a hydrocarbon combustion product shows 8% CO_2, 15% H_2O (vapour), 5.5% O_2 and 71.5% N_2. The combustion product flows steadily through a heat exchanger at 200 kPa pressure. Assume each component in the mixture to be an ideal gas. In order to avoid the condensation of H_2O in the heat exchanger, the minimum allowable temperature (in°C) is ______.

Saturated H_2O Table:

P (kPa)	10	20	30	40	50
T (°C)	45.83	60.09	69.12	75.82	81.35

Solution:

The mass of CO_2 present in 1 m³ of combustion products,

$$(m)_{CO_2}\big|_V = 0.08 \times \frac{44}{22400} \text{kg} = 1.57 \times 10^{-4} \text{ kg}$$

The mass of H_2O present in 1 m³ of combustion products,

$$(m)_{H_2O}\big|_V = 0.15 \times \frac{18}{22400} \text{ kg} = 1.21 \times 10^{-4} \text{ kg}$$

The mass of O_2 present in 1 m³ of combustion products,

$$(m)_{O_2}\big|_V = 0.55 \times \frac{32}{22400} \text{ kg} = 0.79 \times 10^{-4} \text{ kg}$$

The mass of N_2 present in 1 m³ of combustion products,

$$(m)_{N_2}\big|_V = 0.715 \times \frac{28}{22400} \text{ kg} = 8.94 \times 10^{-4} \text{ kg}$$

The number of moles of CO_2 present in 1 m³ of combustion products,

$$(n)_{CO_2}\big|_V = \frac{1.57 \times 10^{-4}}{44} = 3.57 \times 10^{-6}$$

The number of moles of H_2O present in 1 m³ of combustion products,

$$(n)_{H_2O}\big|_V = \frac{1.21 \times 10^{-4}}{18} = 6.72 \times 10^{-6}$$

The number of moles of O_2 present in 1 m³ of combustion products,

$$(n)_{O_2}\big|_V = \frac{0.79 \times 10^{-4}}{32} = 2.47 \times 10^{-6}$$

The number of moles of N_2 present in 1 m^3 of combustion products,

$$(n)_{N_2}\Big|_V = \frac{8.94\times10^{-4}}{28} = 31.93\times10^{-6}$$

The mole fraction of CO_2 in the combustion products,

$$(X)_{CO_2} = \frac{3.57\times10^{-6}}{(3.57+6.72+2.47+31.93)\times10^{-6}} = 0.08$$

The mole fraction of H_2O in the combustion products,

$$(X)_{H_2O} = \frac{6.72\times10^{-6}}{(3.57+6.72+2.47+31.93)\times10^{-6}} = 0.15$$

The mole fraction of O_2 in the combustion products,

$$(X)_{O_2} = \frac{2.47\times10^{-6}}{(3.57+6.72+2.47+31.93)\times10^{-6}} = 0.055$$

The mole fraction of N_2 in the combustion products,

$$(X)_{N_2} = \frac{31.93\times10^{-6}}{(3.57+6.72+2.47+31.93)\times10^{-6}} = 0.715$$

Therefore, the partial pressure of H_2O,

$$P_{H2O} = 0.15 \times 2200 \text{ kPa} = 30\text{kPa}$$

From the given data table, the minimum allowable temperature (in) is 69.12°C.

Hence, answer is 69.12.

Q13. An equimolar mixture of two ideal gases (A, B) expands isentropically in a nozzle. The gas mixture enters the nozzle at 300 kPa, 400 K and exits at 100 kPa. Assuming the mixture to be an ideal gas, the exit temperature of the gas mixture (in K) is _________.

	Molar mass (kg/kmol)	C_p (kJ/kg-K)
Gas A	28.013	1.04
Gas B	2.016	14.21

Solution:

The molar mass of the gas mixture,

$$\mathcal{M}_{mixture} = M_A X_A + M_B X_B = (28.013 \times 0.5 + 2.016 \times 0.5) \text{ kg/kmol} = 15.015 \text{ kg/kmol}$$

Mass of gas A present in 1 mole of the mixture, $m_A|_{mol} = 0.5 \times 28.013 \text{ kg} = 14.0065 \text{ kg}$

Mass of gas B present in 1 mole of the mixture, $m_B|_{mol} = 0.5 \times 2.016 \text{ kg} = 1.008 \text{ kg}$

Therefore, the mass of gas A present in 1 kg of the mixture,

$$m_A\Big|_{mass} = \frac{14.0065}{14.0065+1.008}\ \text{kg} = 0.9329\ \text{kg}$$

And the mass gas B present in 1 kg of the mixture,

$$m_B\Big|_{mass} = \frac{1.008}{14.0065+1.008}\ \text{kg} = 0.0671\ \text{kg}$$

Hence, the specific heat of the mixture

$$\left(C_p\right)_{mixture} = \frac{m_A\big|_{mass}\times\left(C_p\right)_A + m_B\big|_{mass}\times\left(C_p\right)_B}{m_A\big|_{mass} + m_B\big|_{mass}} = 1.9237\ \text{kJ/kg. K}$$

Therefore, the molar specific heat at constant pressure of the mixture,

$$(C_p)_{mixture}\big|_{molar} = (C_p)_{mixture}\times M_{misture} = 28.884\ \text{kJ/kmol. K}$$

Hence, the molar specific heat at constant volume of the mixture,

$$\left(C_v\right)_{mixture}\Big|_{molar}$$

$$= \left(C_p\right)_{mixture}\Big|_{molar} - R = (28.884 - 8.314)\ \text{kJ/kmol. K} = 20.57\ \text{kJ/kmol. K}$$

Therefore, the adiabatic index, $\gamma = \dfrac{\left(C_p\right)_{mixture}\Big|_{molar}}{\left(C_v\right)_{mixture}\Big|_{molar}} = \dfrac{28.884}{20.57} = 1.4$

Now, for an isentropic process of an ideal gas we have

$$\frac{T_2}{T_1} = \left(\frac{p_2}{p_1}\right)^{\frac{\gamma-1}{\gamma}}$$

$$T_2 = T_1\left(\frac{p_2}{p_1}\right)^{\frac{\gamma-1}{\gamma}} = 400\times\left(\frac{100}{300}\right)^{\frac{1.4-1}{1.4}}\ \text{K} = 292.3\ \text{K}$$

Hence, answer is 292.3.

Q14. A rigid vessel of volume 10 m³ is filled with hydrogen at 25 °C and 500 kPa. Due to leakage, some gas has escaped from the vessel until the pressure in the vessel drops down to 200 kPa, and the corresponding temperature of the gas inside the vessel is found to be 15 °C. The amount of gas leaked (in kg) from the vessel is ________.

Solution:

$V = 10\ m^3$

$T_i = 25°\ C$

$P_i = 500\ kPa$

$P_iV_i = n_iRT_i$

$$500 \times 10 = \frac{m_i}{2} \times 8.314 \times 288$$

$m_i = 4.036\ kg$

$P_f = 200\ kPa$

$T_f = 15°C = 288K$

Now, $P_f V = n_f RT_f$

$$200 \times 10 = \frac{m_f}{2} \times 8.314 \times 288$$

$m_f = 1.670$

$\therefore m_i - m_f = (4.036 - 1.670) kg = 2.365 kg$

Hence, the answer is 2.365.

Q15. A hot ideal gas (C_p = 1.2 kJ/(kg.K)) steadily flows through a turbine with inlet and exit temperatures of 1500 K and 500 K respectively. The minimum mass flow rate (in kg/s) of the hot gas to achieve a power output of 12 MW is ____________.

Solution:

$C_p = 1.2$ kJ/kg k

$T_i = 1500$ K

$T_f = 500$ K

$P = 12$ MW

$$\dot{m}h_1 = \dot{m}h_2 + P$$

$$\dot{m} = \frac{P}{h_1 - h_2}$$

$$\dot{m} = \frac{P}{C_p(T_1 - T_2)}$$

$$= \frac{12 \times 10^6}{1.2 \times 10^3 \times (1500 - 500)} \text{ kg/s}$$

$$= 10 \text{ kg/s}$$

Hence, the answer is 10

Q16. Air pressure inside a spherical balloon is proportional to its diameter. The balloon undergoes a reversible, isothermal, non-flow process. During the process, the balloon maintains its spherical shape, and the air inside the balloon consumes 2 kJ of heat. Initial air pressure inside the balloon was 120 kPa, while the initial balloon diameter was 20 cm. Assuming air to be an ideal gas, the final diameter of the balloon (in cm) is ____________.

Solution:

$P \propto D$

Ideal gas, isothermal, non-flow process

$$\delta Q = dU + \delta W$$

$$Q = \int_1^2 pdV \text{ (Since, dU = 0)}$$

$$2\times10^3 = \int_1^2 pdV$$

$$2\times10^3 = \int_1^2 \frac{p_1}{D_1}\times D\times\frac{\pi D^2}{2}dD$$

$$2\times10^3 = \frac{p_1\pi}{2D_1}\times\int_1^2 D^3 dD$$

$$2\times10^3 = \frac{120\times10^3\times3.14}{2\times0.2}\times\left[\frac{D^4}{4}\right]_{D_1}^{D_2}$$

$$2 = \frac{120\times3.14}{2\times0.2}\times\frac{D_2^4 - D_1^4}{4}$$

$$0.00212 = \frac{D_2^4 - (0.2)^4}{4}$$

$$\therefore D_2^4 = 0.1008$$

$$D_2 = 0.3168 \text{ m} = 31.6 \text{ cm}$$

Hence, the answer is 31.6.

Q17. An air-standard diesel engine has a compression ratio of 18 (the ratio of the volume at the beginning of the compression process to that at the end of the compression process), and a cut-off ratio of 2 (the ratio of the volume at the end of the heat addition process to that at the beginning of the heat addition process). The thermal efficiency (in %) of the engine is __________.

Solution:

Diesel cycle

Compression ratio = 18 (r_v)

Cut-off ratio = 2 (r_c)

$$\eta = 1 - \frac{1}{(r_v)^{r-1}} \times \frac{1(r_c)^{r-1}}{r \times r_c - 1}$$

$$\eta = 1 - \frac{1}{(18)^{1.4-1}} \times \frac{1(2)^{1.4-1}}{1.4 \times (2-1)}$$

$$\eta = 1 - \frac{1}{(18)^{0.4}} \times \frac{(2)^{1.4-1}}{1.4 \times (2-1)}$$

$$= 0.63157 = 63.187\ \%$$

Hence, the answer is 63.187.

Q18. Compressed air, at 1 MPa pressure, 400 K temperature flows through a large pipe. An evacuated, insulated rigid tank of 0.5 m³ volume is connected to the pipe through a valve. The valve is opened to fill the tank and the valve closes automatically when the tank pressure reaches 1MPa. Assuming ideal gas behavior, the final air temperature in the tank (in K) is __________.

Solution:

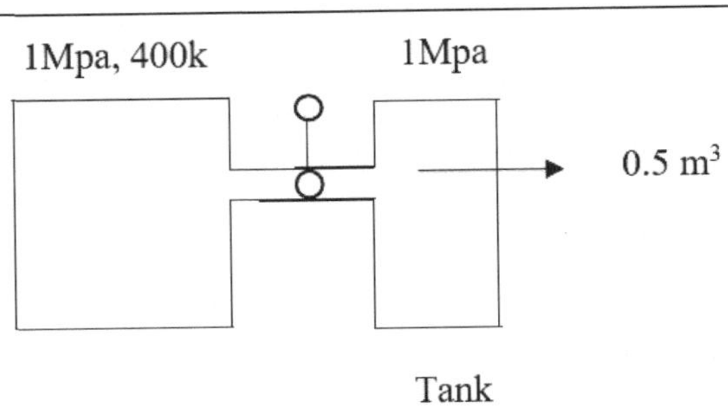

$h_i = u_i + RT_i = u_2$

$C_v T_2 = C_v + RT_i = C_p T_i$

$$T_2 = \frac{C_v + R}{C_v} T_1 = r\, T_1$$

$T_2 = 1.4 \times 400\text{K} = 560$ K

Hence, the answer is 560.

Q19. A 40 kg metal block (C_p = 0.5 kJ/(kg.K)) at T = 450 °C is quenched in 150 kg oil (C_p = 2.5 kJ/(kg. K)) at T = 25°C. If the combined (metal block and oil) system is fully isolated from its surroundings, then the net change in the entropy (in kJ/K) of the combined system is _____________.

Solution:

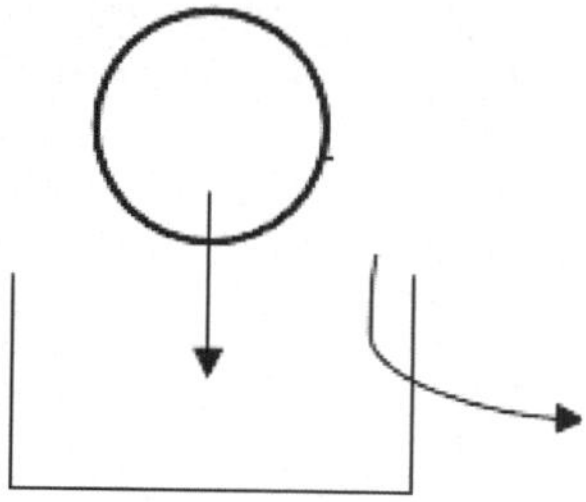

M_2 = 150kg
C_{p2}= 2.5kJ/ kgK
T_2 = 25°C

Metal black
M_1 =40 kg
C_{p1}=0.5kJ/kgK
T_1=450°C

$$\Delta S_1 = m_1 c_{11} \ln \frac{T_f}{T_1}$$

$$\Delta S_2 = m_2 c_2 \ln \frac{T_f}{T_2}$$

$$S_{univ} = \Delta S_1 + \Delta S_2$$

$$= m_1 c_1 \ln \frac{T_f}{T_1} m_2 c_2 \ln \frac{T_f}{T_2}$$

$$m_1 c_1 (t_1 - t_f) = m_2 c_2 (t_f - t_2)$$

$$t_f = \frac{m_1 c_1 t_1 + m_2 + c_2 t_2}{m_1 c_1 + m_2 c_2} = \frac{40 \times 0.5 \times 450 + 150 \times 2.5 \times 25}{40 \times 0.5 + 150 \times 2.5}$$

t_f = 46.51°C = 319.51 K

$$\text{Change of entropy} = 40 \times 0.5 \times \ln\left(\frac{319.51}{723}\right) + 150 \times 2.5 \ln\left(\frac{319.51}{298}\right)$$

= −16.3324+26.13 = 9.80

Hence, the answer is 9.80.

Q20. For phase change from solid (sol) to liquid (liq) state, if the slope of the solid-liquid coexistence line in the *P-T* diagram is negative, then:

(A) $v_{liq} < v_{sol}$ (B) $v_{liq} > v_{sol}$

(C) $s_{liq} < s_{sol}$ (D) $h_{liq} < h_{sol}$

Solution:

Hence, the answer is (A).

Q21. A house-hold refrigerator operates under steady state condition between an evaporator temperature of 263 K and a condenser temperature of 323 K. The heat load to the refrigerator is 3 kW. The actual COP of the refrigerator is half of that of a Carnot refrigerator operating between the same condenser and evaporator temperatures. The power required (in kW) to run the refrigerator is __________.

Solution:

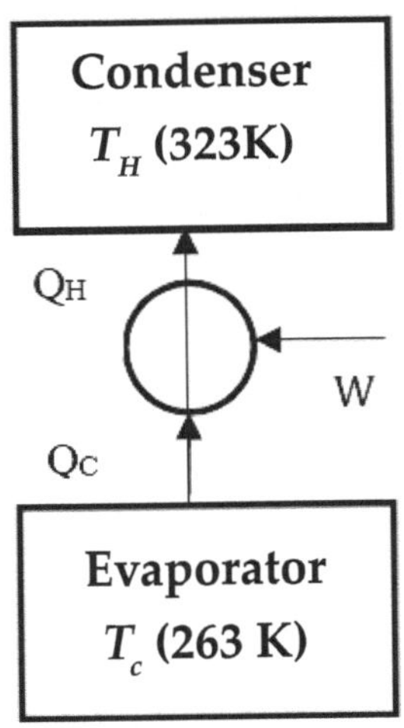

$$COP = \frac{T_L}{T_H - T_L}$$

$$COP = \frac{263}{323 - 263} = 4.3833$$

$$\text{Actual COP} = \frac{4.3833}{2} = 2.1916$$

$$2.1916 = \frac{Q_2}{W} = \frac{3}{W}$$

$W = 1.36$ kW

$P = 1.36$ kW

Hence, answer is 1.36.

Q22. The Maxwell relation that results from the expression for the Helmholtz free energy $A = U - TS$, is:

(A) $\left.\frac{\partial T}{\partial v}\right)_s = -\left.\frac{\partial P}{\partial s}\right)_v$

(B) $\left.\frac{\partial T}{\partial P}\right)_S = \left.\frac{\partial v}{\partial s}\right)_P$

(C) $\left.\frac{\partial P}{\partial T}\right)_v = \left.\frac{\partial s}{\partial v}\right)_T$

(D) $\left.\frac{\partial v}{\partial T}\right)_P = -\left.\frac{\partial s}{\partial P}\right)_T$

Solution:

$A = U\text{-}TS$

$dA = dU - TdS - SdT$

$dA = \text{-}Pdv - sdT$

$$-\left(\frac{\partial p}{\partial T}\right)_v = -\left(\frac{\partial s}{\partial v}\right)_T$$

Hence, the answer is (C).

GATE 2015
THERMODYNAMICS

Q. 1 – Q. 9 carry one mark each.

Q1. **A gas expands following the relation PV^n = constant, from the initial state P_1, V_1 to final volume $V_2 = 2V_1$. For the values of 'n' mentioned below, maximum displacement work is obtained for**

(A) $n = -1$ (B) $n = 0$

(C) $n = 1$ (D) $n = 1.4$

Solution:

Given:

PV^n= constant

Initial state= P_1V_1

Final state =P_2V_2

$V_2 = V_1$

For polytropic process,

$$P_1V_1^n = P_2V_2^n$$

$$\frac{P_1}{P_1} = \left(\frac{V_2}{V_1}\right)^n = \left(\frac{2V_1}{V_1}\right)^n$$

$$\frac{P_1}{P_2} = 2^n$$

$$\therefore P_2 = \frac{P_1}{2^n}$$

Work done in polytropic process $(n \neq 1)$ is given by,

$$w = \frac{P_1V_1 - P_2V_2}{n-1}$$

$$= \frac{P_1V_1 - (P_2 \times 2V_1)}{n-1}$$

$$=\frac{P_1V_1\left\{1-\frac{1}{2^n}\times 2\right\}}{n-1}$$

$$=P_1V_1\left\{\frac{1-2^{1-n}}{n-1}\right\}$$

$$w=\frac{P_1V_1}{n-1}-\frac{P_1V_1\left(2^{1-n}\right)}{n-1}$$

For an isothermal process, $n = 1$ and $PV = P_1V_1 = P_2V_2 =$ Constant $= K$ (say)

The work done in an isothermal process is given by,

$$W_{n=1}=\int_{V1}^{V2}PdV$$

$$=\int_{V_1}^{V_2}\frac{K}{V}dV$$

$$=K\ln\left(\frac{V_2}{V_1}\right)$$

$$=P_1V_1\ln\left(\frac{2V_1}{V_1}\right)$$

$$=P_1V_1\ln(2)$$

$= 0.6931\ P_1V_1$

For the different values of n, work done is calculated by the above formula,

***n*=0**	***n*=-1**	***n*=1.4**
$W=P_1V_1\left\{\frac{1-2^1}{-1}\right\}$ $=P_1V_1$	$W=P_1V_1\left\{\frac{1-2^2}{-2}\right\}$ $=1.5\ P_1V_1$	$W=P_1V_1\left\{\frac{1-2^{1-1.4}}{1.4-1}\right\}$ $=0.6054\ P_1V_1$

∴The maximum displacement work corresponds to w = 1.5 P_1V_1, for which n = -1.

Hence, the answer is (A).

Q2. A 100 Ω electrical resistor is heated steadily by passing a current of 20 A. If heating is performed in the ambient at 30°C, the rate of increase in entropy of the universe in kW/K is ______.

Solution:

Given:

$R = 100$

$I = 20\text{A}$

$T = (30 + 273)\text{K} = 303\text{ K}$

From the 2nd law of Thermodynamics,

$$\Delta S = \frac{dQ}{T} = \frac{I^2R}{T} = \frac{(20)^2 \times 100}{303} = \frac{4 \times 10^4}{303} \times 10^3\,\text{W/K} = 0.132\text{ kW/K}$$

$= 0.132$ kW/K

Hence, the answer is 0.132 kW/K.

Q3. As per Clausius inequality, a system operating on an irreversible cycle transfers

(A) more entropy to the sink than what it receives from the source

(B) as much entropy to the sink as it receives from the source

(C) less entropy to the sink than what it receives from the source

(D) less entropy to the sink than that corresponding to a reversible cycle

Solution:

According to the Clausius inequality,

For irreversible process, $ds > \dfrac{dQ}{T}$

$\therefore S_2 - S_1 > \dfrac{dQ}{T}$

∴ System operating on irreversible cycle transfers more entropy to the sink than, what it receives from the source.

Hence, the answer is (A).

Q4. The critical point of a substance corresponds to the state

(A) At which the solid, liquid and vapour phases are in equilibrium

(B) Beyond which liquid will require very large amount of heat to become vapour

(C) Beyond which solid sublimates directly to the vapour state

(D) Beyond which the distinction between liquid and vapour phases disappears

Solution:

Critical point (or *critical* state) is the end *point* of a phase equilibrium curve. At critical point, defined by critical temperature and pressure, phase boundaries between liquid and vapor vanish, i.e. it is the point beyond which no distinct liquid and vapour phases exist.

Hence, the answer is (D).

Q5. Consider the process of sensible cooling of air with 60% relative humidity at constant pressure. Which one of the following statements is TRUE for this process?

(A) Both humidity ratio and relative humidity increase during the process.

(B) The humidity ratio decreases continuously due to condensation.

(C) The dry bulb temperature decreases but the wet bulb temperature increases.

(D) The humidity ratio remains constant.

Solution:

Sensible cooling of air with 60% RH,

Pressure = constant

During sensible cooling dry bulb temperature decreases but as there is no change in the specific humidity.

Also, its wet bulb temperature decreases, enthalpy decreases, RH increases.

Hence, the answer is (D).

Q6. The coefficient of performance (COP) of a reversible refrigerator operating between two thermal reservoirs is 4.0. The efficiency (in percentage) of a reversible heat engine operating between the same temperature limits is ______.

Solution:

Given:

$$COP_{ref} = 4$$

$$COP_{ref} = \frac{T_2}{T_1 - T_2}$$

$$4 = \frac{T_2}{T_1 - T_2}$$

$$\therefore \frac{T_1}{T_2} = \frac{5}{4}$$

The efficiency of reversible H.E operating between the same limits is:

$$\eta_{H.E.} = 1 - \frac{T_2}{T_1}$$

$$=1-\frac{4}{5}$$

$= 0.2$

Hence, answer is 20%.

Q7. Differential change in specific enthalpy of a superheated vapor for which ideal gas approximation does not hold, is given by the expression

(A) $dh = C_p\, dT$

(B) $dh = C_p dT + \left.\frac{\partial h}{\partial v}\right|_T dv$

(C) $dh = C_p dT + \left.\frac{\partial h}{\partial p}\right|_T dp$

(D) $dh = C_v dT + \left.\frac{\partial h}{\partial p}\right|_T dp$

Solution:

$dh = Tds + vdp.$

From the 2nd Tds equation

$$Tds = C_p dT - T\left(\frac{\partial v}{\partial T}\right)_p dp$$

$$\therefore dh = C_p dT + \left\{v - T\left(\frac{\partial v}{\partial T}\right)_p\right\} dp$$

$$v - T\left(\frac{\partial v}{\partial T}\right)_p = \left.\frac{\partial h}{\partial p}\right|_T$$

Since, an ideal gas is a function of temperature only, hence $\left.\frac{\partial v}{\partial p}\right|_T = 0$.
i.e., for an ideal gas, $dh = C_p dT$
But, for a superheated vapour that cannot be approximated as an ideal gas,

$$dh = C_p dT + \left\{v - T\left(\frac{\partial v}{\partial T}\right)_p\right\} dp = C_p dT + \left.\frac{\partial h}{\partial p}\right|_T.$$

Hence, the answer is (C).

Q8. An ideal gas mixture of oxygen (molecular weight = 32 kg/kmol) and carbon dioxide (molecular weight = 44 kg/kmol) has a mass composition of 40% and 60% respectively. If the total pressure is 200 kPa, the partial pressure of oxygen (in kPa) is ______.

Solution:

Mass composition of $O_2 = 40\%$

$\therefore (m)_{O_2} = 40g$

Mass composition of $CO_2 = 60\%$

$\therefore (m)_{CO_2} = 60g$

Mole fraction of $O_2, (X)_{O_2} = \frac{\text{no. of moles of } O_2}{\text{total moles}}$

$$(X)_{O_2} = \frac{40/32}{\frac{40}{30} + \frac{60}{44}}$$

Using Dalton's law of Partial pressure,

(Partial pressure) $O_2 = (X)_{O_2} \times P_{total}$

$= 0.4783 \times 200$ kPa

$= 95.66$ kPa

Hence, the answer is 95.66 kPa.

Q9. In an ideal Rankine cycle, increase in superheat of vapor at the exit of boiler leads to

(A) decrease in net work output from the cycle

(B) increase in cycle efficiency

(C) decrease in cycle efficiency

(D) decrease in quality of steam at the exit of the turbine

Solution:

In rankine cycle, increase in superheat at the boiler exit increases the η as:

$$\eta_{Rankine} = 1 - \frac{T_2}{T_m}$$

$T_m \rightarrow$ mean temp of heat addition, which is a constant temperature, such that:
$\Delta h = T_m \Delta s$, where Δh & Δs are the enthalpy difference, during the process of heat addition.

From eq^n (i), on increasing T_m, η of Rankine cycle increases.

Hence, the answer is (B).

Q. 10 – Q. 22 carry two marks each.

Q10. Two moles of air at 1 atm, 21.1°C goes through an adiabatic device and separates into a hot stream of 0.4 moles at 1 atm, 176.3°C and a cold stream of 1.6 moles at 1 atm, –17.7°C, without any external work. It can be concluded that

(A) the total entropy change is zero

(B) the total entropy change for the process is positive

(C) the device violates Second Law of Thermodynamics

(D) the device violates First Law of Thermodynamics

Solution:

Two moles of air at 1 atm, 21.1°C, separates into a hot stream of 0.4 moles at 1 atm, 176.3°C and a cold stream of 1.6 moles at 1 atm, –17.7°C, without any external work, so there is transfer of heat from a low temperature to a high temperature, without any transfer of work, which violates the second law of thermodynamics.

Hence, the answer is (C).

Q11. For a real gas undergoing volume expansion through a porous plug with $\alpha = \frac{1}{v}\frac{\partial v}{\partial T}\Big|_p$ the Joule -Thomson cooling effect is observed if

(A) $0 < \alpha T < 1$ (B) $\alpha T = 1$ (C) $\alpha T > 1$ (D) $\alpha T = 0$

Solution:

Joule thomson cooling effect is observed if,

$\mu_j > 0$. (The region inside the inversion curve, where is positive, is the cooling region).

From the 2nd Tds eq^n,

$$Tds = c_p dT - T\left(\frac{\partial v}{\partial T}\right)_p dp$$

$$dh = Tds + vdp$$

$$dh = c_p dT - \left[T\left(\frac{\partial v}{\partial T}\right)_p - v\right]dp$$

$$\mu_j = \left(\frac{\partial T}{\partial P}\right)_h$$

$$\therefore \mu_j = \frac{1}{C_P}\left\{T\left(\frac{\partial v}{\partial T}\right)_p - v\right\}$$

$$\mu_j = 1\Big/C_p(\alpha Tv - v)\left(\text{Given, } \alpha = \frac{1}{v}\frac{\partial v}{\partial t}\Big|_p\right)$$

$\mu_j = v/C_p(\alpha T - 1)$

For $\mu_j > 0$,

$\alpha T = 1 > 0$

$\alpha T > 1$

Hence, the answer is (C).

Q12. A lead bullet at 100°C travelling at 500 m/s, strikes a target and adiabatically comes to rest. If the specific heat of lead is 92J/kg°C, melting temperature is 327.5°C and heat of fusion is 108kJ/kg, the percentage mass of lead in the bullet that is melted by collision is ______.

Solution:

From S.F.E.E, $Q = \Delta E_K + \Delta E_P + \Delta U + W$

$\Delta E_P + \Delta U + W = 0$

Total heat energy (due to melting & latent heat) = change in kinetic energy

m_L = mass of lead

m_T = Total mass of bullet

$$m_L \times C_p \times \Delta T + m_L \times L = \frac{1}{2} \times m_T \times (v_1^2 - v_2^2) \quad (v_2 = 0)$$

$$m_L \times 92 \times (327.5 - 100) + m_L \times 108 \times 10^3 = 1/2 \times m_T \times (500)^2$$

$$m_L (92 \times 227.5 + 10800) = \frac{1}{2} \times m_T \times 25 \times 10^4$$

$$\therefore \frac{m_L}{m_T} = \frac{25 \times 10^4}{2 \times (92 \times 227.5 + 108000)} = 0.96951$$

$$\% \left(\frac{m_L}{m_T} \right) = 96.951\%$$

Hence, the answer is 96.951.

Q13. An air-water vapour mixture with volume 100 m³ at 100 kPa, 35°C is at 75% relative humidity. Taking saturation pressure of vapour at 35°C as 5.63 kPa, the mass of vapour in the mixture in kg is ______.

Solution:

Given:

$V = 100\ m^3$

$P = 100$ kPa,

$T = 35°C=308$ K

R.H. = 75%,

P_S (at 35°C) = 5.63 kPa

$$RH = \frac{Pv}{Ps} = m_v / m_{v(sat)}$$

$$0.74 = \frac{p_v}{5.63}$$

$P_v = 4.225$ kPa

Hence, partial pressure of dry air, $P_a = (100\text{-}4.225)$ kPa = 95.775 kPa

Let the mass of dry air in the given volume be m_a.

Then $P_a V = m_a RT$

$$\therefore m_a = \frac{P_a V}{RT} = \frac{957.775 \times 100}{0.287 \times 308} \text{kg} = 108.35 \text{ kg}$$

$\omega = 0.622 \times \left(\frac{P_v}{P_a}\right)$, where ω = Humidity Ratio = mass of vapour/mass of dry air

$$0.622 \times \left(\frac{4.225}{100 \times 4.225}\right) = \frac{mv}{108.35}.$$

$$0.622 \times \left(\frac{4.225}{100 \times 4.225}\right) = \frac{mv}{108.35}$$

$m_v = 2.97$ kg

Hence, the answer is 2.97

Q14. An ideal gas of 1 kg mass enclosed inside a rigid vessel at the initial temperature 1200 K is employed as a heat source. The specific heat C_v of the gas is 718 J/kg K. The maximum work in kJ that can be developed by operating a heat engine between the ideal gas and the ambient at 300 K is

(A) 646.2 (B) 484.7

(C) 387.7 (D) 347.6

Solution:

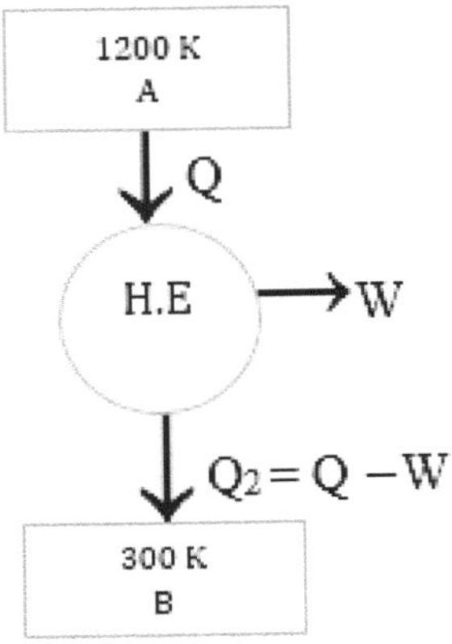

Given:

$m = 1$ kg,

$C_v = 718$ J/kgK

By the entropy principle,

$\Delta S_{univ} \geq 0.$

$\Delta S_A + \Delta S_B \geq 0.$

Heat lost by the ideal gas = $mC_V \Delta T$

$Q = 1 \times 0.718 \times (1200\text{-}300)$

$Q = 0.718 \times 900$ kJ

$$\Delta S_A = mC_v \ln\left(\frac{T_0}{T_1}\right)$$

where, T_1 and T_0 are the temperature of ideal gas and ambient air, respectively.

$$\Delta S_B = \frac{Q - W}{T_0}$$

$$\Delta S_A + \Delta S_B \geq 0$$

$$mC_v \ln\frac{T_o}{T_1} + \frac{Q - W}{300} \geq 0$$

For, maximum work (to be done by H.E,

$$mC_v \ln\frac{T_0}{T_1}+\frac{Q}{300}\geq\frac{W}{300}$$

$$300\times\left\{mC_v \ln\frac{T_0}{T_1}\right\}+Q=W_{max}$$

$$300\times\left\{1\times 0.715\times\ln\frac{300}{1200}\right\}+0.718\times 900=W_{max}$$

$$W_{max}=347.6 \text{ kJ}$$

Hence, the answer is (D).

Q15. The boiling point of water changes from 99.62 to 105.99 when the system pressure is changed from 1 bar to 1.25 bar. The boiling point value (in °C) corresponding to a pressure of 1.5 bar is ________.

Solution:

Given:

P_1= 1 bar

P_2 = 1.25 bar

T_1 = 99.62°C= 372.62 K

T_2 = 105.99°C= 378.99 K

Let the enthalpy of vaporization be ΔH_{vap}. Then the Clausius-Clapeyron equation gives

$$\ln\left(\frac{P_1}{P_2}\right)=\frac{\Delta H_{vap}}{R}\left(\frac{1}{T_2}-\frac{1}{T_1}\right)$$

$$\frac{\Delta H_{vap}}{R}\left(\frac{1}{378.99}-\frac{1}{372.62}\right)=\ln\left(\frac{1}{1.25}\right)$$

$$\frac{\Delta H_{vap}}{R}=4946.96 \qquad \text{(i)}$$

Let the boiling point at P_3 = 1.5 bar be T_3.

Hence, the Clausius-Clapeyron equation gives

$$\ln\left(\frac{P_1}{P_3}\right)=\frac{\Delta H_{vap}}{R}\left(\frac{1}{T_3}-\frac{1}{T_1}\right)$$

$$4946.96\times\left(\frac{1}{T_3}-\frac{1}{372.62}\right)=\ln\left(\frac{1}{1.5}\right) \text{[From eq. (i)]}$$

$$\frac{1}{T_3} = \frac{1}{372.62} - \frac{0.04054}{4946.96}$$

$$T_3 = 383.36 \text{ K} = 111.36°\text{C}$$

Hence, the answer is 111.36

Q16. An ideal gas expands in an adiabatic frictionless nozzle from the inlet conditions of 31 bar, 800 K to the ambient pressure of 1 bar at the outlet. The specific heat C_p for the gas is 1 kJ/kg K and the specific heat ratio γ = 1.4. Neglecting inlet kinetic energy, the velocity of the gas (in m/s) at the nozzle exit is

(A) 32 (B) 500 (C) 707 (D) 1000

Solution:

P_1 = 31 bar

T_1 = 800K

P_2 = 1 bar

C_p = 1kJ/kgK

Applying the condition of reversible adiabatic process,

$$\frac{T_2}{T_1} = \left(\frac{P_2}{P_1}\right)^{\frac{\gamma-1}{\gamma}}$$

$$\frac{T_2}{800} = \left(\frac{1}{31}\right)^{0.4/1.4}$$

$\therefore T_2 = 299.907$

Applying the SFEE eq^n for nozzle,

$$\therefore mh_1 + mv_1^2 = mh_2 + mv_2^2$$

$$v_2 = \sqrt{2(h_1 - h_2)}$$

$$v_2 = \sqrt{2(c_p(T_1 - T_2))}$$

$$v_2 = \sqrt{2\times10^3 \times (800 - 299.907)}$$

$v_2 = 1000.0929$

Hence, the answer is (D).

Q17. One kmol of hydrogen (molecular weight = 2 kg/kmol, specific heat ratio γ = 1.4) at 1 bar, 300 K mixes with one kmol of nitrogen (molecular weight = 28 kg/kmol, specific heat ratio γ = 1.4) at 1 bar, 300 K in an adiabatic vessel. The final mixture is also at 1 bar, 300 K. The entropy change (in kJ/K) for the process is

(A) – 5.76 (B) zero

(C) 5.76 (D) 11.53

Solution:

Given:

No. of moles of hydrogen, (n_{H_2}) =1

No. of moles of nitrogen, (n_{N_2}) =1

For both Hydrogen, Nitrogen,

P = 1 bar.

T = 300K

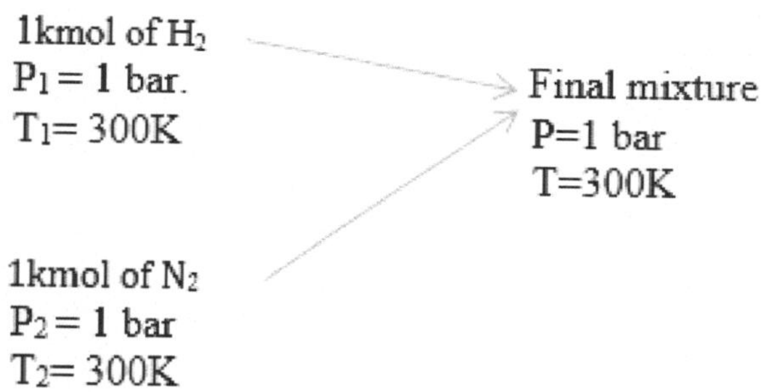

Using the Gibb's theory of entropy change, due to diffusion,

$\Delta S = -R\,\{n_1 \ln x_1 + n_2 \ln x_2\}$,

where, x_1 and x_2 are the mole fraction of hydrogen and nitrogen respectively,

$= -8.314 \times \{1 \times \ln(0.5) + 1 \times \ln(0.5)\}$

$= -2 \times 8.314 \times \ln(0.5) = 11.53$

Hence, the answer is (D).

Q18. A cycle 1-2-3-1 is proposed with the following processes:

1-2: Constant pressure expansion, 2-3: Reversible adiabatic expansion, 3-1: Irreversible adiabatic compression. Which one of the following statements is TRUE?

(A) The net work of the cycle is zero because there is no heat transfer.

(B) The cycle is feasible and can deliver net positive work.

(C) The cycle is impossible according to Kelvin-Planck statement.

(D) The cycle is impossible in accordance with First Law of Thermodynamics.

Solution:

1-2: Constant pressure expansion, so there is heat addition.

2-3: Reversible adiabatic expansion, so work is done by the system.

2-3: Irreversible adiabatic compression, so work is done on the system.

This cycle produces net work by exchanging heat with only one reservoir which is impossible according to Kelvin-Planck statement.

Hence, the answer is (C).

Q19. One kg of saturated liquid-vapour mixture of water at 150 kPa (u_f = 467 kJ/kg, v_f = 0.001053 m³/kg; u_g = 2520 kJ/kg and v_g = 1.159 m³/kg) with quality of 0.7 is enclosed in a piston cylinder assembly. Heat is added at constant pressure to this system while a paddle wheel transfers a work of 50 kJ. The mixture eventually attains saturated vapour state. The amount of heat added to the mixture (in kJ) is ______.

Solution:

Given:

$m = 1$ kg

$P = 150$ kPa

$u_f = 467$ kJ/Kg

$v_f = 0.001053$ m³/kg

$u_g = 2520$ kJ/kg

$v_g = 1.159$ m³/kg

$x = 0.7$

$w = -50$ kJ

$u_1 = u_f + x.u_{fg}$

$= 467 + 0.7\{2520\text{-}467\}$

$=1904.1$ kJ/kg

$\Delta u = u_2 - u_1 = (2520 - 1904.1)$ kJ/kg $= 615.9$ kJ/kg

Initial specific volume of the liquid-vapor mixture

$\therefore v_1 = v_f + x.v_{fg}$

$= 0.001053 + 0.7 \times (1.159 - 0.001053)$ m³/kg $= 0.811$ m³/kg

$v_2 = v_g = 1.159$ m³/kg (Final state is saturated vapor state)

$v = v_2 - v_1 = (1.159\text{-}0.811)$ m³/kg$= 0.348$ m³/kg

$\therefore \Delta H = m(\Delta u + P\Delta v)$ (As, heat is added at constant pressure)

$= 1 \times \{(615.9) + 150 \times 0.348\}$ kJ $= 668.1$ kJ

$\therefore Q = \Delta H + W$ (as W is the work, which is transferred to the gas through a paddle wheel)

$= (668.1 - 50)$ kJ $= 618.1$ kJ

Hence, the answer is 618.1.

Q20. Air enters a pipe at 1 bar and flows isothermally at the rate of 1 kg/s. Due to pipe friction, the pressure drop between two sections of the pipe is 7% of the pressure at inlet section. For ambient temperature, T_0 = 300 K, the rate of irreversibility (in W) between the two sections is ______.

Solution:

Given:

$P_1 = 1$ bar,

$\dot{m} = 1$ kg/s

$T_0 = 300$K

$P_1 - P_2 = 7\% \times P_1$

$$P_1 - P_2 = \frac{7}{100} \times P_1$$

$$P_1 \left\{1 - \frac{7}{100}\right\} = P_2$$

$\therefore P_2 = 0.93P_1$

From the 2nd law of thermodynamics, the net entropy change, exceeds the entropy transfer, as that due to the entropy generation, which arises due to intrinsic irreversibility.

$$\Delta s_{gen} = C_p \ln\left(\frac{T_2}{T_1}\right) - R\ln\left(\frac{P_2}{P_1}\right)$$

$\therefore$ Rate of irreversibility $(\dot{I}_0) = \dot{m}T_0\ \Delta S_{gen}$

$$= 1 \times 300\left\{C_p \ln\left(\frac{T_2}{T_1}\right) - R\ln\left(\frac{P_2}{P_1}\right)\right\}$$

$$= 300 \times \left\{-\frac{8.314}{29} \times \ln\left(\frac{0.93}{1}\right)\right\} \times 10^3\text{W},$$ as process is isothermal, $T_2 = T_1$

$\dot{I}_0 = 6241.5$W

Hence, the answer is 6241.5.

Q21. A pump raises pressure of saturated liquid water at 100 kPa (density ρ = 959 kg/m³) to 2MPa. The isentropic efficiency of the pump is 0.92. The work done by the pump (in J/kg) is _____.

Solution:

Given,

$P_1 = 100$KPa,

Density (ρ) = 959 kg/m^3

P_2 = 2MPa

Isentropic efficiency $\eta = 0.92$

Work done by the pump is given by:

$w_{ideal} = v \times (P_1 - P_2) \times 10^5$

$$= \frac{1}{959} \times (20-1) \times 10^2 \text{ J/kg}$$

= 1981.230 J/kg

$\eta = 0.92$

Isentropic η (of pump) = w_{ideal}/ w_{actual}

$$\therefore w_{actual} = \frac{w_{ideal}}{\eta}$$

$$= \frac{1981.23}{0.92} \text{J/kg} = 2153.51 \text{ J/kg}$$

Hence, the answer is 2153.51 J/kg.

Q22. In a Brayton cycle with air (γ = 1.4) as working fluid, T_{min} = 300 K and T_{max} = 1000 K. The pressure ratio corresponding to maximum net work per cycle is _______.

Solution:

Given,

γ =1.4

T_{min} = 300 K

T_{max} = 1000 K

The work capacity of the Brayton cycle, operating between T_{max} and T_{min} , is zero, when, r_p = 1 and when r_p=0. There, is an optimum value of pressure ratio (r_p), where, the work is maximum.

Corresponding to maximum work, the pressure ratio in Brayton cycle is given by:

$$= \left(\frac{T_{max}}{T_{min}}\right)^{\frac{1.4}{2\times 0.4}}$$

$$= \left(\frac{1000}{300}\right)^{\frac{1.4}{2\times 0.4}} = 8.2231$$

Hence, the answer is 8.223.

GATE 2014
THERMODYNAMICS

Q. 1 – Q. 9 carry one mark each.

Q1. Entropy is a

(A) Path function (B) Point function

(C) Property independent function (D) Neither path nor point function

Solution:

Entropy is a point function as it is a property of the system, and for a given state, there is a definite value for each property, irrespective of the path followed and depends only on the initial and final states.

Hence, the answer is (B).

Q2. A small container has gas at high pressure. It is placed in an evacuated space. If the container is punctured, work done by the gas is

(A) Positive (B) Negative (C) Zero (D) ∞

Solution:

When the the container is punctured, the gas in the container expands in the evacuated space, until the pressure inside the container and space becomes equal. This represents the case of free expansion, and the work done by the gas, becomes zero.

Hence, the answer is (C).

Q3. The molecular weight of a mixture is 38.4 gm/mol. The mixture is composed of methane and carbondioxide gases. The atomic weights of the elements C, H, and O are 12, 1, and 16 gm/mole, respectively. The mole fraction of methane ($X_{methane}$) is __________ and that of carbon-dioxide ($X_{carbon\text{-}dioxide}$) is ______.

(A) $X_{methane} = 0.2; X_{carbondioxide} = 0.8$

(B) $X_{methane} = 0.8; X_{carbondioxide} = 0.2$

(C) $X_{methane} = 0.3; X_{carbondioxide} = 0.7$

(D) $X_{methane} = 0.7; X_{carbondioxide} = 0.3$

Solution:

Molecular weight of $CH_4 (M_1) = 16$ gm/mol

Molecular weight of $CO_2 (M_2) = 44$ gm/mol

Molecular weight of Mixture (M_{mixt}) =38.4 gm/mol

Also,

$M_{mixt} = M_1x_1 + M_2x_2$, where $x_1$and x_2 are the mole fractions of CH_4 and CO_2 respectively.

$38.4 = 16x_1 + 44x_2$

$16x_1 + 44(1\text{-}x_1) = 38.4$

$16x_1 + 44 - 44x_1 = 38.4$

$-28x_1 = -5.6$

Mole fraction of CH_4 $(x_1) = 0.2$

Mole fraction of CO_2 $(x_2) = 0.8$

Hence, the answer is (A).

Q4. A system undergoes a change from state 1 to state 2. During this process, the change in the internal energy is ΔU. The change in internal energy of the system when executing the cycle 1-2-1 is equal to

(A) ΔU (B) 2ΔU (C) Zero (D) -2ΔU

Solution:

Change in internal energy, from state 1 to state 2 =ΔU

Change in internal energy, from state 2 to state 1= -ΔU

Change in internal energy during the cycle (1-2-1) = ΔU + (-ΔU) = 0

Hence, the answer is (C).

Q5. Which among the following plots represents a line joining two states with the same dew point temperature on a standard psychometric chart, with the dry bulb temperature on the X-axis and the humidity ratio on the Y-axis?

(A)

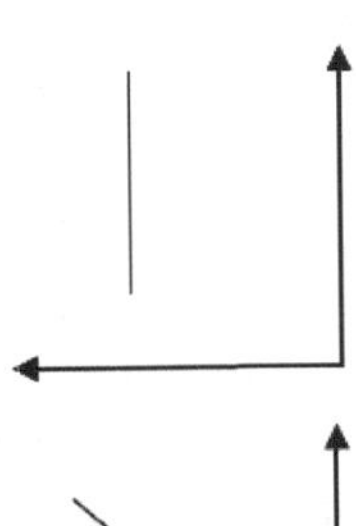

(B)

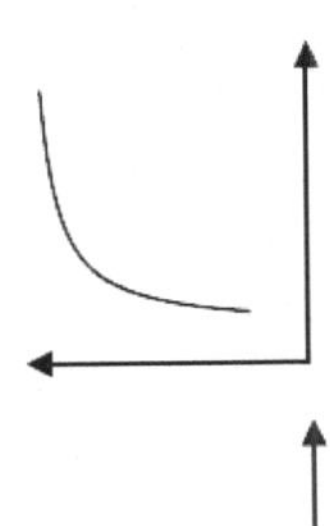

(C)

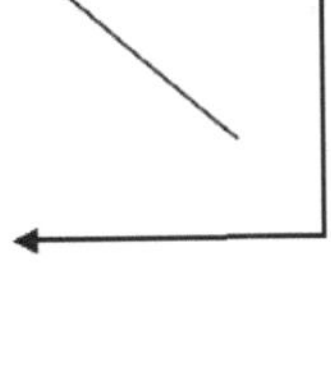

(D)

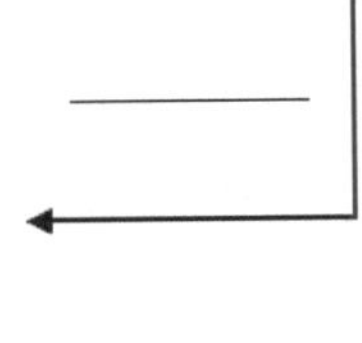

Solution:

Dew point line are oriented parallel to the humidity ratio lines. Therefore, a line joining two states with the same dew point temperature on a standard psychrometric chart, will be a horizontal line, parallel to the x- axis.

Hence, the answer is (D).

Q6. The efficiency of a reversible engine operating between two temperatures is 40%. The COP of a reversible refrigerator operating between the same temperatures is

(A) 1.5 (B) 2.5 (C) 0.4 (D) 3.5

Solution:

η of HE = 40%

$$\eta_{HE} = \frac{T_H - T_L}{T_H} = 0.4$$

where T_H and T_L are the source and sink temperatures respectively.

$$T_H - T_L = 0.4T_H$$

$$\frac{T_L}{T_H} = 0.6$$

$$\text{COP (operating betwen the same temperatures)} = \frac{T_L}{T_H - T_L}$$

$$= \frac{T_L}{0.4T_H} = 0.6/0.4 = 1.5$$

Hence, the answer is (A).

Q7. For a superheated vapor that cannot be approximated as an ideal gas, the expression determining a small change in the specific internal energy is

(A) $du = C_p\, dT + \left.\frac{\partial u}{\partial v}\right|_T dv$ (B) $du = C_p\, dT + \left.\frac{\partial u}{\partial P}\right|_T dv$

(C) $du = C_v\, dT + \left.\frac{\partial u}{\partial v}\right|_T dv$ (D) $du = C_v dT$

Solution:

From the 1st Tds relation

$$Tds = C_v dT + T\left(\frac{\partial P}{\partial T}\right)_v dv$$

$$du = Tds - Pdv$$

$$\therefore du = C_v dT + \left\{T\left(\frac{\partial P}{\partial T}\right)_v - P\right\} dv$$

$$T\left(\frac{\partial P}{\partial T}\right)_v - P = \left.\frac{\partial u}{\partial v}\right|_T$$

Since an ideal gas is a function of temperature only, $\therefore \left.\frac{\partial u}{\partial v}\right|_T = 0$, for an ideal gas.
i.e., for ideal gas, $du = C_v\, dT$
However, for a superheated vapor that cannot be approximated as an ideal gas,

$$du = C_v dt + \left\{T\left(\frac{\partial P}{\partial T}\right)_v - P\right\} dv$$

$$= C_v dT + \left.\frac{\partial u}{\partial v}\right|_T$$

Hence, the answer is (C).

Q8. The minimum and maximum volumes in an air standard Otto cycle are 100 and 800 cm³. Its thermal efficiency (%) is

(A) 56.47 (B) 94.55 (C) 54.08 (D) 87.50

Solution:

For an Otto cycle,

$$r_v \text{ (volumetric compression ratio} = \frac{v_{\max}}{v_{\min}} = \frac{800}{100} = 8$$

$$\eta(\text{Eficiency}) = 1 - \frac{1}{r_v^{\gamma-1}} = 1 - \frac{1}{8^{1.4-1}} = .5647$$

$\eta = 56.47\%$

Hence, the answer is (A).

Q9. At a saturation temperature T_{sat}, the difference between the entropy of saturated vapor and entropy of saturated liquid can be expressed as

(A) $(h_f - h_g)/T_{sat}$ (B) $(h_g - h_f)/T_{sat}$
(C) $(u_g - h_f)/T_{sat}$ (D) $(u_f - h_g)/T_{sat}$

Solution:

From 1st and 2nd law of thermodynamics,

$Tds = dh - vdp$

$$\left(\frac{\partial h}{\partial s}\right)_p = T_{sat}$$

(change of phase from gas to liquid occurs at constant pressure at the saturation temperature)

Also, from the Mollier diagram or *h-s* plot,s the slope of an isobar on the *h-s* coordinates is equal to the absolute saturation temperature.

$\therefore ds = \dfrac{h_g - h_f}{T_{sat}}$

Hence, the answer is (B).

Q. 10 – Q. 22 carry two marks each.

Q10. A gas in a closed system is compressed reversibly from an initial volume of 0.2 m^3 to 0.1 m^3 at a constant pressure of 3 bar. During this process, there was a heat transfer of 50 kJ from the gas. The change in internal energy of the gas during this process in kJ is

(A) 20 (B) -80 (C) 80 (D) -20

Solution:

Given:

V_1 (initial volume) = 0.2 m^3

V_2 (final volume) = 0.1 m^3

Pressure (constant) = 3 bar =3×10^2 kPa

Q = -50kJ (-ve sign, due to heat removal/transfer from the gas)

From the 1st law of Thermodynamics,

$Q = \Delta U + W$

$-50 = \Delta U + p\Delta V$

$\Delta U + (0.1-0.2)\times3\times10^2 = -50$

$\Delta U = -20$ kJ

Hence, the answer is (D).

Q11. In a closed rigid vessel, air is initially at a pressure of 0.3 MPa and volume of 0.1 m^3 at 300 K. A stirrer supplies 100 kJ of work to the air, while 20 kJ of heat is lost to the atmosphere across the container walls. After these processes, the temperature of air changes to ________K.

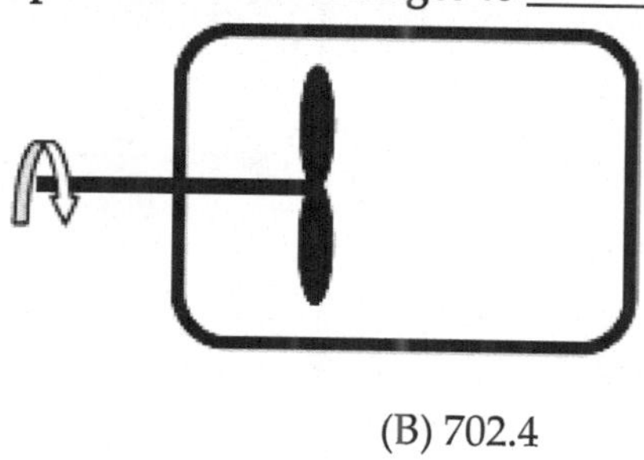

(A) 321.9 (B) 702.4

(C) 782.4 (D) 620.2

Solution:

Given:

p_1 = 0.3 MPa

$V_1 = 0.1$ m^3

$T_1 = 300$ K

$Q = -20$kJ (-ve, since heat is lost to the atmosphere)

$W = -100$ kJ (-ve, since work is done on the system)

From, the 1st law of Thermodynamics,

$Q = \Delta U + W$

$\Delta U = Q - W$

$= \{-20-(-100)\}$ kJ

$\Delta U = 80$ kJ (i)

From ideal gas law,

$p_1V_1 = mRT_1$

$0.3\times1000\times0.1 = m\times0.287\times300$

$m = 0.348$ kg (ii)

Now,

$\Delta U = mC_v\Delta T$

$80 = 0.348 \times 0.717 \times (T_2-300)$ [from eqs. (i) and (ii)]

$T_2 = 620.6$ K

Hence, the answer is (D).

Q12. A reversible heat engine (E) operates using three thermal reservoirs with temperatures as shown in the following figure. If Q_1=Q_2, the efficiency of the engine is ___________.

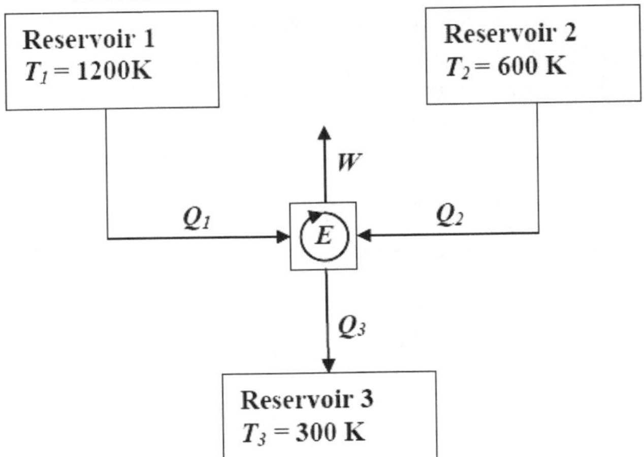

(A) 0.25 (B) 0.125

(C) 0.625 (D) 0.75

Solution:

$$\eta = \frac{\text{Work output}}{\text{Heat input}} = \frac{W}{Q_1 + Q_2} = \frac{Q_1 + Q_2 - Q_3}{2Q_1} = 1 - \frac{Q_3}{2Q_1}$$

$$Q_1/T_1 + Q_2/T_2 = Q_3/T_3 \left(\text{For a reversible3 cycle, } \oint \frac{\delta Q}{T} 0 \right)$$

Q_1/ 1200+ Q_2/600 = Q_3/300

Q_1/ 1200+ Q_1/600 = Q_3/300 (Given, $Q_1 = Q_2$)

$$\therefore Q_3 = Q_1 \times 300 \left(\frac{1}{1200} + \frac{1}{600} \right)$$

$$\therefore Q_3 = \frac{3Q_1}{4}$$

$$\eta = 1 - \frac{Q_3}{2Q_1}$$

= 1-(3/8) (substituting the value of Q_3 from eq^n (i)

= 0.625

Hence, the answer is (C).

Q13. A metal block of mass 25 kg at 300 K is immersed in an infinitely large liquid nitrogen bath maintained at 77 K. The system comprising of the block and liquid nitrogen attains thermal equilibrium. The average specific heat of the metal is 0.45 kJ/(kg.K). The entropy generated during the process is __________kJ/K.

(A) 17.28 (B) 32.5 (C) 47.8 (D) -47.8

Solution:

Given:

Mass of the metal block (m) = 25 kg

Temperature of the metal block, T_1 = 300 K,

Specific heat of the metal block, (C_p) = 0.45 kJ/kg k

$$(S_1 - S_1)_{block} = mC_p \ln\left(\frac{T_2}{T_1}\right) - mR \ln\left(\frac{V_2}{V_1}\right)$$

$$= 25 \times 0.45 \times \ln\left(\frac{77}{300}\right) \text{ kJ/K}$$

(Since the metal block is immersed in an infinitely large liquid nitrogen bath, so initial and final volume are same)

$(S_2 - S_1)_{block} = -15.299$ kJ/K

$Q = mC_p \Delta T$

$Q = 25 \times 0.45 \times (300 - 77)$ kJ

$Q = 2508.75$ kJ

Therefore, the change in entropy of the liquid nitrogen

$$(S_2 - S_1)_{\text{nitrogen}} = \frac{Q}{T_{nitrogen}}$$

From the 2nd law of Thermodynamics, entropy change exceeds the entropy transfer, on account of entropy generation, which is due to internal irreversibility in the process.

$$\therefore S_2 - S_1 = \frac{Q}{T_{nitrogen}} + \Delta S_{generated}$$

$$-15.299 = -2508.75/77 + \Delta S_{generated}$$

$$\Delta S_{generated} = 17.2 \text{ kJ/K}$$

Hence, the answer is (A).

Q14. For a gas obeying the equation of state given by $\left(P + \frac{a}{v^2}\right)v = RT$ the values of the critical volume and the critical temperature are 0.004 m³/kg and 100 °C, respectively. If the value of the gas constant is 250 J/(kg.K), then the value of the constant '*a*' is ____________ (N.m⁴/kg²). Note that the critical point is the point of inflection on the critical isotherm.

(A) 124.3 (B) 0.75

(C) 186.58 (D) 248.67

Solution:

As the critical point is the point of inflection on critical isotherm,

$$\therefore \left(\frac{\partial P}{\partial v}\right)_{T=Tc} = 0 \qquad \text{(i)}$$

Given:

$$\left(P + \frac{a}{v^2}\right)v = RT \qquad \text{(ii)}$$

From eq. (i), differentiating P w.r.t v at constant temperature, we get,

$$P = \frac{a}{v^2}$$

Substituting the value of P in eq. (ii), we get,

$$\frac{2a}{v} = RT$$

$\therefore 2a = 250 \times 373 \times 0.004$

$a = 186.5$

Hence, answer is (C).

Q15. A rigid closed vessel is initially filled with 2 kg of water which is a mixture of saturated liquid and saturated vapor states at 2 bar. The vessel is placed in an oven which heats the mixture to the critical state. Using the saturated and critical property values from the table given below, the heat transferred from the oven to the vessel is __________kJ.

Pressure = 2 bar			
v_f(m³/kg)	v_g (m³/kg)	u_f(kJ/kg)	u_g (kJ/kg)
0.0010605	0.8857	504.49	2529.5
Critical pressure			
	v_c (m³/kg)	u_c (kJ/kg)	
	0.003155	2029.6	

(A) 3035.8 (B) 3040.6 (C) 3036.2 (D) 3044.9

Solution:

Given:

At $P = 2$ bar,

$v_f = 0.0010605$ m³/kg

$v_g = 0.8857$ m³/kg

$u_f = 504.49$ kJ/kg

$u_g = 2529.5$ kJ/kg

$u_c = 2029.6$ kJ/kg

At critical state, $v_c = 0.003155$ m³/kg

As the vessel is rigid, therefore the volume remains the same.

i.e., $v = v_c$

$v = v_f + x.v_{fg} = v_c$

$\therefore v_c = 0.0010605 + x(0.8857 - 0.0010605)$

$0.003155 = 0.0010605 + x \times (0.8857 - 0.0010605)$

$x = 0.002367$

$u = u_f + x.u_{fg}$

$= 504.49 + 0.002367 \times (2529.5 - 504.49)$

$= 509.2831$ kJ/kg

Change in internal energy per unit mass $= (u_c - u)$

$= (2029.6 - 509.2831)$ kJ/kg

$= 1520.3$ kJ/kg

Change in internal energy = 1520.3 × 2 kJ (mass = 2 kg)
= 3040.6 kJ

Hence, the answer is (B).

Q16. The equation of state for a certain gas is given by $v = RT / P - C_1 / T^2 + C_2$, where C_1 is 50,000 ($K^2.m^3$)/kg and C_2 is 0.8 m^3/kg. The relation $\frac{\partial h}{\partial P}\Big|_T = v - T\frac{\partial v}{\partial T}\Big|_P$ is known for the gas. The inversion temperature, given by the condition, $\left[\frac{\partial v}{\partial P}\Big|_T = 0\right]$ is ________K.

(A) 500.0 (B) 433.0 (C) 353.6 (D) 250.0

Solution:

Given:

$$v = \frac{RT}{P} - \frac{C_1}{T_2} + C_2$$

Inversion temperature is the temperature at which gas shows neither cooling effect nor heating effect, i.e., at this temperature, Joule-Thomson coefficient $\mu_j = 1$
The inversion temperature, given by the condition,

$$\left[\frac{\partial v}{\partial P}\Big|_T = 0\right]$$

$$\left(\frac{\partial h}{\partial P}\right)_T = v - T\left(\frac{\partial v}{\partial T}\right)_P \quad \text{(given)}$$

$$\left(\frac{\partial v}{\partial T}\right)_P = \frac{R}{P} + \frac{2C_1}{T^3}$$

$$T\left(\frac{\partial v}{\partial T}\right)_p = \frac{RT}{P} + 2C_1/T^2$$

$$P = v - T\left(\frac{\partial v}{\partial T}\right)_P$$

$$\frac{RT}{P} - \frac{C_1}{T^2} + C_2 - \frac{RT}{P} - \frac{2C_1}{T^2}$$

$$= -\frac{C_1}{T^2} - \frac{2C_1}{T^2}TC_2$$

$$-\frac{3C_1}{T^2} + C_2 = 0$$

$$-3C_1 + T^2C_2 = 0$$

$$-3 \times 50000 + T^2 \times 0.8 = 0$$

$$T^2 = \frac{150000}{0.8}$$

$$T = 433.0 \text{ K}$$

Hence, the answer is (B).

Q17. The maximum pressure and temperature in an air standard diesel cycle are 44 bar and 1600 K, respectively. If the minimum pressure and temperature are 1 bar and 300 K, respectively, then the cut-off ratio (the ratio of the volume at the end of the heat addition process to that at the beginning of the heat addition process) is

(A) 1.000 (B) 14.920 (C) 2.809 (D) 1.809

Solution:

P_2= 44 bar, P_1= 1 bar

T_3 (Max. temp) = 1600 K

T_1 (Min. temp) = 300K

Applying the condition of is entropic compression, for process (1-2) in diesel cycle,

$$\frac{T_2}{T_1} = \left(\frac{P_2}{P_1}\right)^{\frac{\gamma-1}{\gamma}}$$

T_2= 2.95 × (300) = 884.46 K

For the process (2 – 3) in diesel cycle, P = const.

$$\therefore V_2/T_2 = V_3/T_3$$

$V_3/V_2 = T_3/T_2$ = 1600/884.46 = 1.809

Hence, the answer is (D).

Q18. The thermal efficiency of an air standard Brayton cycle 0.35. The pressure ratio across the turbine is

(A) 4.516 (B) 5.232 (C) 7.535 (D) 8.234

Solution:

$\eta_{Brayton} = 1 - \frac{1}{(r_p)^{\frac{\gamma-1}{\gamma}}}$, where r_p is the pressure ratio, which is the ratio of maximum and minimum pressure, i.e. pressure limits on which Brayton cycle works.

$$\eta_{Brayton} = 1 - \frac{1}{(r_p)^{\frac{\gamma-1}{\gamma}}} = 0.35$$

$\therefore r_p = 4.516$

Hence, the answer is (A).

Q19. Steam is isentropically expanded in a turbine from 80 bar to 7 bar. At the inlet of the turbine (state 1) h_1 is 3246 kJ/kg and s_1 is 6.52 kJ/(kg.K).

Pressure = 7 bar			
h_f (kJ/kg)	h_g (kJ/kg)	s_f [kJ/(kg.K)]	s_g [kJ/(kg.K)]
697	2763	2.0	6.7

The enthalpy of the steam exiting the turbine (state 2) in kJ/kg is

(A) 2683.87 (B) 2657.17 (C) 1986.87 (D) 3354.17

Solution:

Given:

$P_1 = 80$ bar

$P_2 = 7$ bar

$h_1 = 3246$ kJ/kg

$s_1 = 6.52$ kJ/kg K

$h_f = 697$ kJ/kg

$h_g = 2763$ kJ/kg

Applying the condition of reversible isentropic expansion in turbine,

$s_1 = s_2$

$s_2 = s_f + xs_{fg}$

$s_1 = s_2 = 2 + x \times (6.7-2)$

$6.52 = 2 + x \times 4.7$

$x = 0.96$

$h_2 = h_f + xh_{fg} = \{697 + 0.96 \times (2763-697)\}$ kJ/kg $= 2680.36$ kJ/kg

Hence, the answer is (A).

Q20. A thin insulating membrane separates two tanks initially filled with nitrogen [mean C_v = 21.6 J/(mol.K)] and carbondioxide [mean C_v = 11.6 J/(mol.K)] as shown below.

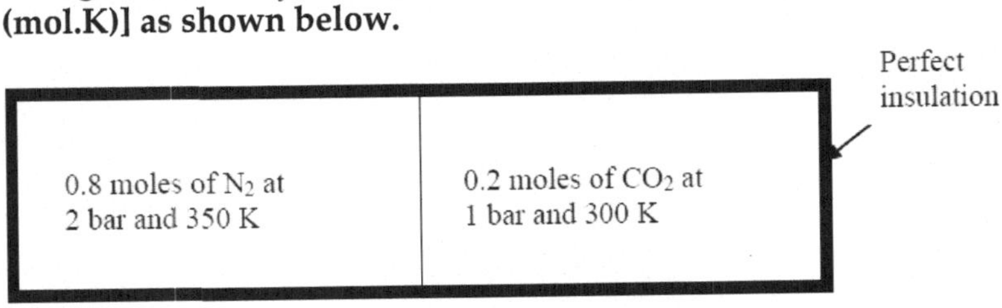

The membrane is ruptured and the gases are allowed to mix to form homogeneous mixture at equilibrium. During this process there are no heat

or work interactions between the tank contents and the surroundings. The final temperature at the equilibrium state in Kelvin is

(A) 344.1 (B) 306.3 (C) 325.0 (D) 346.1

Solution:

Given:

For $N_{2,}$

$n_1 = 0.8$

$m_1 = 0.8 \times 28$ gm

$C_{v1} = 21.6$ J/(mol.K)

$T_1 = 350$ K

For $CO_{2,}$

$n_2 = 0.2$

$m_2 = 0.2 \times 44$ gm

$C_{v2} = 11.6$ J/(mol.K)

$T_2 = 300$ K

As, there are no heat interactions between the tank contents and the surroundings and also the work transfer is zero.

$\therefore Q = 0$ and $W = 0$.

Therefore

$n_1 Cv_1 \Delta T_1 = n_2 C_{v2} \Delta T_2$

$n_1 Cv_1 (T_1 - T_f) = n_1 Cv_1 (T_f - T_1)$

$$T_f = \frac{n_1 C_{v1} T_1 + n_2 C_{v2} T_2}{n_1 C_{v1} + n_2 C_{v1}}$$

$T_f = 344.1$ K

Hence, the answer is (A).

Q21. Two moist air streams MAS[1] and MAS[2] are mixed adiabatically. The details of MAS[1] and MAS[2] are given below in the table.

	MAS[1]	MAS[2]
h (kJ/kg of dry air)	42	80
V (m^3/kg of dry air)	0.85	0.9
Flow rate (m^3/min)	85	90

With pressure remaining same and with no work interactions during the mixing process, the enthalpy of the mixed stream is __________ kJ/kg of dry air.

(A) 122 (B) 61 (C) 81 (D) 108

Solution:

For, MAS_1

Flow rate (m^3/min) = 85

v (m^3/kg dry air) = 0.85

$$\therefore \text{Mass flow rate of dry air } = \frac{\frac{85}{0.85}\text{kg}}{\text{min}} = 100 \text{ kg/min}$$

Similarly, for MAS_2, mass flow rate of dry air = 100 kg/min

From, conservation of mass,

MAS_1+MAS_2 = Mixed steam

Hence, the mass flow rate of the mixed stream = 200 kg/min

$$\dot{m}_1 h_1 + \dot{m}_2 h_2 = \dot{m}_3 h_3$$

$100 \times 42 + 100 \times 80 = 200 \times h_3$

$h_3 = 61$

Hence, the answer is (B).

Q22. Consider the steady flow of air through an insulated nozzle. The pressure and temperature at the inlet are 120 kPa and 320 K, respectively. The outlet pressure is 1 bar. The inlet velocity is very small and the air undergoes a reversible adiabatic process. The outlet velocity, in m/s, is

(A) 303.7 (B) 180.7 (C) 5.7 (D) 127.3

Solution:

Given:

P_1 = 120 kPa

T_1 = 320K

Applying the condition of reversible adiabatic process,

$$\frac{T_2}{T_1} = \left(\frac{P_2}{P_2}\right)^{(\gamma-1)/\gamma}$$

$$\frac{T_2}{320} = \left(\frac{100}{120}\right)^{(1.4-1)/1.4}$$

$$\frac{T_2}{320} = (0.84)^{0.28}$$

T_2 = 303.34 K

$$\Delta h = C_p \Delta T = 1.005 \times (320 - 304.7) = 15.38 \times 10^3 \text{ J}$$

Applying the SFEE eq^n for nozzle, $V = \sqrt{2\Delta h}$

$V = 175.38$ m/s ≈180 m/s

Hence, the answer is (B).

GATE 2013
THERMODYNAMICS

Q. 1 – Q. 9 carry one mark each.

Q1. The measured temperature of a system is 30°C. Its exact absolute temperature in K is

(A) 303.00 (B) 303.10 (C) 303.15 (D) 303.16

Solution:

Given:

$T = 30°C$

T in (K) = T °C +273.15

= 30 + 273.15

= 303.15 K

Hence, the answer is (C).

Q2. The fuel air mixture in a pertrol engine is ignited with a spark plug at the end of compression stroke. This process

(A) increases the entropy of the fuel air mixture but decreases the entropy of the spark plug

(B) decreases the entropy of the fuel air mixture but increases the entropy of the spark plug

(C) decreases the entropy of the fuel air mixture and of the spark plug

(D) increases the entropy of the fuel air mixture and of the spark plug

Solution:

When the fuel air mixture is ignited with a spark plug, there is an increase in pressure and temperature, thereby causing an increase in the entropy of the fuel air mixture and the spark plug.

Hence, the answer is (D).

Q3. In the van der Waals equation of state given below

$$\left(p+\frac{a}{v^2}\right)(v-b)=RT$$

The constant a represents the effect of:

(A) attractive forces between molecules
(B) repulsive forces between molecules
(C) deviation from molecules being spherical
(D) finite size of the molecule

Solution:

The van der Waals law is based on the plausible reasons that real gases deviate from the ideal gas law. The ideal gas law treats gas molecules as point particles, which are neither attracted or repelled. For, real gases, van der Waal provided for intermolecular attraction, by adding to the observed pressure p in the equation of state a term a.

$$\left(p+\frac{a}{v^2}\right)(v-b)=RT$$

In the equation, a represents attractive forces between the molecules.

Hence, the answer is (A).

Q4. For a reversible isothermal expansion of an ideal gas from a state 1 to a state 2,

(A) $s_1 = s_2$ (B) $s_1 > s_2$ (C) $s_1 < s_2$ (D) $h_1 > h_2$

Solution:

For a reversible isothermal expansion, $dT = 0$

$$S_2 - S_1 = C_v \ln\left(\frac{T_2}{T_1}\right) + R\ln\left(\frac{v_2}{v_1}\right)$$

For an expansion process, $v_2 > v_1$

$\therefore S_2 > S_1$

Hence, the answer is (C).

Q5. For a pure substance the critical isotherm on the *p-v* plane exhibits

(A) a maximum (B) a minimum
(C) a point of inflection (D) a discontinuity

Solution:

The critical isotherm at critical state on *p-v* plane, where the three roots of vanderwal's eq^n coincides, has a zero slope but also, its slope changes at the critical state such that:

$$\left(\frac{\partial p}{\partial v}\right)_T = 0$$

$$\left(\frac{\partial^2 p}{\partial v^2}\right)_T = 0$$

Therefore, the critical isotherm has a point of inflection (change of curvature) at the critical point.

Hence, the answer is (C).

Q6. For an ideal gas as a working fluid for a given heat input Q, the process that gives the maximum work among the following four processes is

(A) isothermal (B) constant volume

(C) constant pressure (D) isentropic

Solution:

From the 1st law of thermodynamics,

$Q = \Delta U + W$

$W = Q - \Delta U$

For an isothermal process, $dT = 0$

$W = Q.$

For and isochoric process, $dV = 0.$

As , hence

$W = 0.$

For a constant pressure process,

$W = Q - \Delta U$

$= nC_p\Delta T - nCv\Delta T.$

$= n\,(C_p - Cv_)\,\Delta T.$

For an isentropic process,

$Q = \Delta U + W.$

$T\Delta s = \Delta U + W.$

$W = -\Delta U$

Therefore the work done is the maximum in an isothermal process.

Hence, the answer is (A).

Q7. An air standard Otto cycle has the following shape on a thermodynamic property plane.

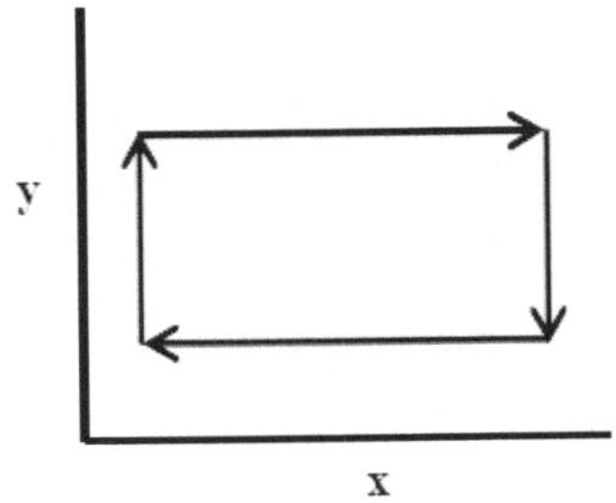

The x and y coordinates, respectively, are

(A) v and p (B) s and v

(C) v and T (D) s and p

Solution:

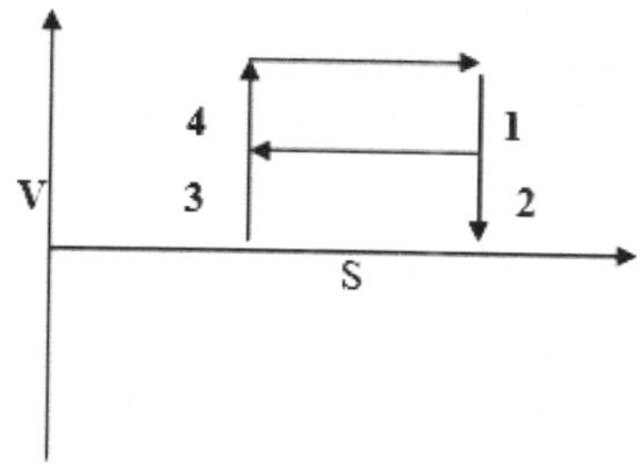

1-2 isentropic expansion,

2-3 Isochoric heat rejection,

3-4 Isentropic compression,

4-1 Isochoric Heat addition

Hence, the answer is (B).

Q8. The specific volume of steam after expansion in a turbine is 12 m³/kg. At this pressure the saturated liquid and saturated vapour specific volumes are 0.001 and 15.25 m³/kg respectively. What is the dryness fraction to second decimal place accuracy?

Solution:

Given:

Specific volume of steam, after expansion (v) = 12m³/kg

Volume of saturated liquid, (v_f) = 0.001m³/kg

Volume of saturated vapor, (v_g) = 15.25 m³/kg

$v = v_f + x.v_{fg}$

12= 0.001 + x(15.25-0.001)

x = 0.786

Therefore, the dryness fraction= 0.79

Hence, the answer is 0.79

Q9. Which of the following processes, shown in the figure below, represents the throttling of an ideal gas?

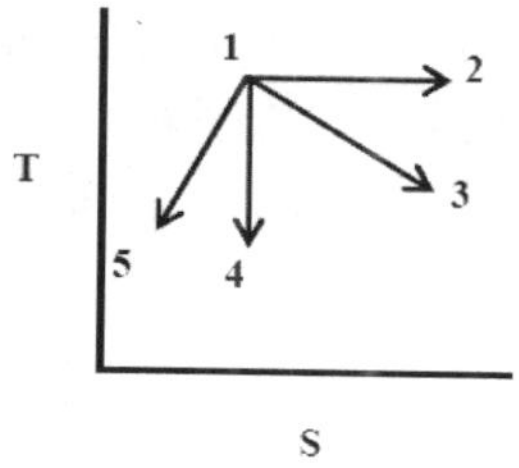

(A) 1 to 2 (B) 1 to 3 (C) 1 to 4 (D) 1 to 5

Solution:

For ideal gas enthalpy is a function of temperature, i.e., $dh = nCpdT$.

And in throttling, the enthalpy remains constant, therefore the temperature remains same. In the graph, the line depicting constant temperature is 1-2

Hence, the answer is (A).

Q. 10 - Q. 22 carry two marks each.

Q10. On a ln(p) vs h coordinate system, where ln(p) is the y-coordinate and h is the x-coordinate, the slope of a constant entropy line is

(A) $1/v$ (B) v (C) p/v (D) $1/(pv)$

Solution:

Given:

ln(p) v/s h coordinate system where,

$\ln(p) = y -$ coordinate

$h = x -$ coordinate

Also, $h = u+pv$

$\therefore dh = du+pdv+vdp$ (i)

From the thermodynamic identity

$Tds = du + pdv$

$Tds = dh - vdp$

Gibbs free energy function $g = h\text{-}Ts$, | $dh = vdp$ (when entropy is constant)

Slope $(m) = dy/dx$

$$= \frac{d\{\ln(p)\}}{dh}$$

$$= \frac{\left(\frac{1}{p}\right)dp}{vdp}$$

$$= \frac{1}{pv}$$

Hence, the answer is (D).

Q11. Starting from the definition of Gibbs free energy function $g = h - Ts$, the Maxwell relation that can be derived is

(A) $\left.\frac{\partial p}{\partial T}\right|_S = \frac{v}{s}$ (B) $\left.\frac{\partial p}{\partial T}\right|_S = v$

(C) $\left.\frac{\partial v}{\partial s}\right|_T = -\left.\frac{\partial p}{\partial T}\right|_S$ (D) $\left.\frac{\partial v}{\partial T}\right|_p = -\left.\frac{\partial s}{\partial p}\right|_T$

Solution:

Given:

Gibbs free energy function, $g = h - Ts$,

On differentiating the function we get,

$dg = dh - Tds - sdT$

$dg = Tds + vdp - Tds - sdT$ (Since, $dh = Tds + vdp$)

$dg = vdp - sdT$

$\therefore \left.\frac{\partial s}{\partial p}\right|_T = -\left.\frac{\partial v}{\partial T}\right|_P$

Hence, the answer is (D).

Q12. A thermodynamic cycle operates between one source at a temperature of 600 K, another source at a temperature of 300 K and a sink at a temperature T as shown in the figure below.

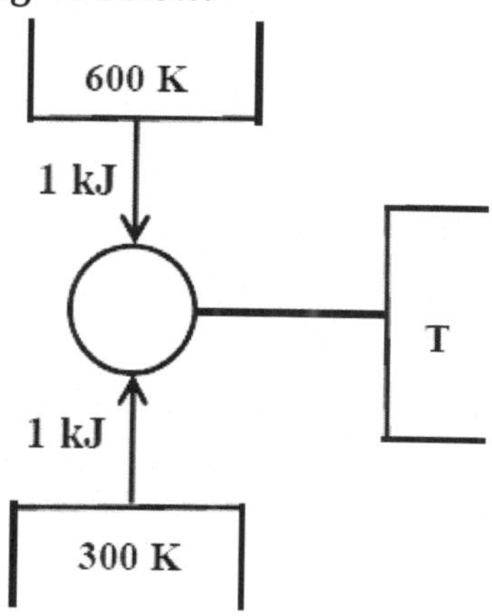

If the First and Second laws of thermodynamics are not violated, what should be the value of T in K?

Solution:

Given:

Source temperature $(T_1) = 600$ K

Source temperature $(T_2) = 300$ K

Sink temperature = T

Heat received by H.E from source at T_1=1 kJ

Heat received by H.E from source at T_2 =1 kJ

Heat transferred to the sink at T be Q_2

According to Clausius theorem, the cyclic integral of $\frac{dQ}{T}$ for of a reversible cycle is zero.

$$\text{i.e., } \oint \frac{dQ}{T} = 0$$

$$\frac{Q_1}{600} + \frac{Q_1}{300} = \frac{Q_2}{T}$$

$$\frac{1}{600} + \frac{1}{300} = \frac{Q_2}{T}$$

$$\frac{1+2}{600} = \frac{Q_2}{T}$$

$$\frac{3}{600} = \frac{2}{T}$$

$$T = \frac{2 \times 600}{300} \text{K}$$

$$T = 400 \text{ K}$$

Hence, answer is 400 K.

Q13. A closed system containing an ideal gas undergoes a cycle as shown in the figure shown below. For the process 1-2, which one of the following statements is true?

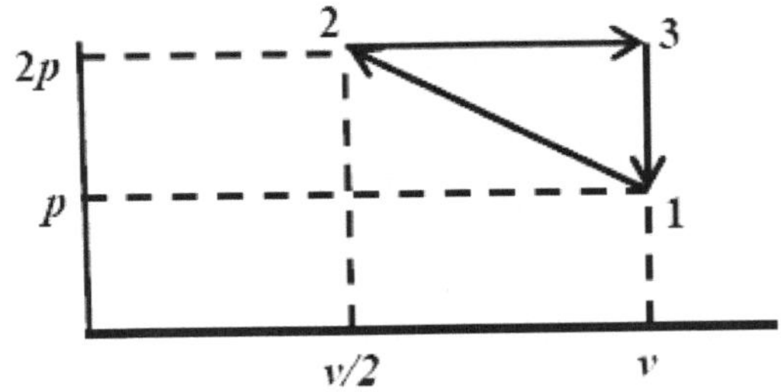

(A) Heat added $= \frac{3}{4} pv$ (B) Heat rejected $= \frac{3}{4} pv$

(C) Heat added $= \frac{1}{2} pv$ (D) Heat rejected $= \frac{1}{4} pv$

Solution:

For the process 1-2,

At the initial state (1), flow energy $= p \times v$

At the final state (2), flow energy $= 2p \times v/2 = pv$.

i.e., $d(pv) = 0$. (as the value of pv remains same)

For an ideal gas,

$pv = nRT$

$d(pv) = nRdT$

$0 = nRdT$

For an ideal gas, the internal energy is a function of temperature only.

$\therefore dU = 0$ (for the process 1-2)

$\delta Q = dU + \delta W$

$\delta Q = \delta W$

From graph, given in the question,

Area of the curve under process (1-2)

$$= \frac{1}{2} \times \frac{v}{2} \times p + \frac{pv}{2}$$

$$= \frac{3vp}{4}$$

Work done for the process $1-2 = -\frac{3vp}{4}$ = (area under p-v curve represents work done)

$\therefore$ Heat is rejected, as volume is decreasing.

Heat rejected $= \frac{3vp}{4}$

Hence, the answer is (B).

Q14. A well-insulated rigid hot water tank receives steady flow of water from two sources as shown in the figure below

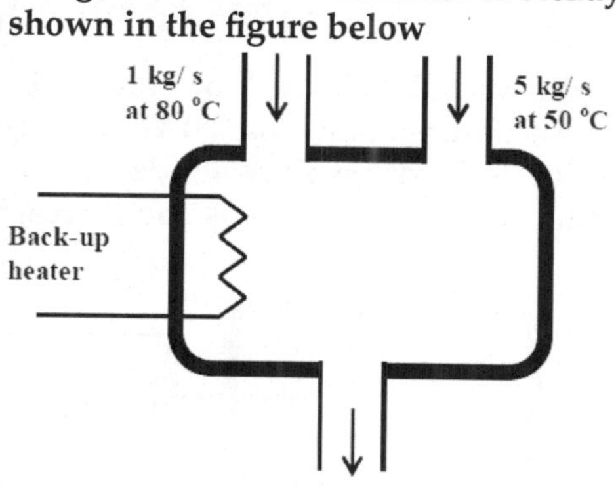

There is no accumulation of water in the tank. A back-up heater is provided to ensure a constant outflow temperature of water at 60°C from the tank under steady state. What is the required capacity of the back-up heater to the nearest kW?

Solution:

Given:

$\dot{m}_1 = 1 \text{ kg/s}$

$\dot{m}_2 = 5 \text{ kg/s}$

where $\dot{m}_1$ and $\dot{m}_2$ are the amount of water entering the tank from the two sources 1 and 2, respectively.

By the law of conservation of mass, if there is no accumulation $\left(\frac{dm}{dt} = 0\right)$ (the net amount of mass flow rate entering the control volume, ($\in_{mi}$) must equal the amount of mass flow rate leaving the control volume ($\in_{me}$).

$$\in_{mi} - \in_{me} = \frac{\partial m}{\partial t}$$

$\in_{mi} - \in_{me}$

$\in_{me} = \dot{m}_1 + \dot{m}_2$

$\in_{me} = 6$ kg/s (given)

Now, applying the law of conservation of energy to the control volume

$$\dot{m}_1 h_1 + \dot{m}_2 h_2 + \dot{Q} = \dot{m}_3 h_3$$

$$\dot{Q} = \dot{m}_3 h_3 - \dot{m}_1 h_1 - \dot{m}_2 h_2$$

$$= 6 \times 4.18 \times 60 - 1 \times 4.18 \times 80 - 5 \times 4.18 \times 50 \text{ kJ/s}$$

$$= 125.4 \text{ kW}$$

Hence, the answer is 125.4.

Q15. 1 kg of air in an insulated rigid tank of volume 1 m^3 is churned with a friction-less fan (see figure below) of 600 W capacity for 10 minutes. The fan efficiency is 100 %. Treating air as an ideal gas and neglecting kinetic and potential energy changes, what is the increase of pressure, to the nearest kPa?

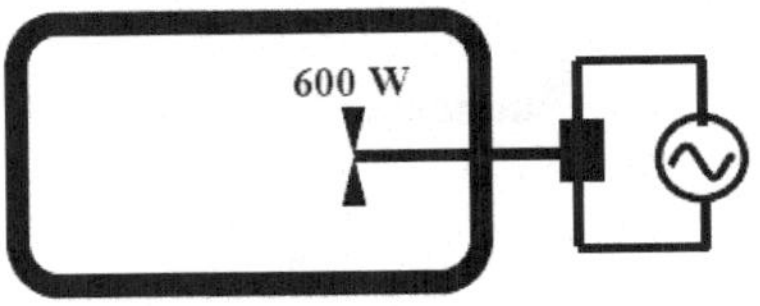

Solution:

Given:

$m = 1$ kg

$V = 1$ m^3

Capacity of fan = 600 W

Efficiency of fan = 100%

For 1 sec, energy obtained = 600 J

When it is churned for 10 mins, work energy supplied= 600 X 600 J =360 kJ.

This means that the work done, W = -360 kJ (Negative, since the work is done on the system)

$Q = \Delta U + W$ (Q=0, since the tank is insulated)

$\Delta U = -W$

$m \times C_V \times \Delta T = 360$

$1 \times 0.718 \times \Delta T = 360$

$\Delta T = 501.39$ K

From ideal gas law,

$PV = mRT$

$V\Delta P = 1 \times 0.287 \times \Delta T$ (Since V is constant)

$\Delta P \times 1 = 0.287 \times 501.39$

$\Delta P = 143.89$ kPa

Hence, the answer is 143.89.

Q16. The isothermal compressibility of a liquid is 5×10^{-6}/kPa. If it is compressed at constant temperature from 5000 to 10000 kPa, what is the ratio of final volume to initial volume, to second decimal place accuracy? __________

Solution:

Given:

Isothermal compressibility, $k_T = 5 \times 10^{-6}\ \text{kPa}^{-1}$

Initial Pressure, P_1 = 5000 kPa

Final Pressure, $P_2 = 10^4$ kPa

$K_T = \frac{-1}{v_1} \times \left(\frac{\partial v}{\partial P}\right)_T$, where v_1, v_2 are the initial and final volume respectively.

$$= \frac{-1}{v_1} \times \frac{(v_2 - v_1)}{(10^4 - 5000)}$$

$$\frac{-v_2}{v_1} + 1 = 5 \times 10^{-6} \times (10^4 - 5000)$$

$$\therefore \frac{v_2}{v_1} = 0.975$$

The ratio of final volume to initial volume is 0.975.

Hence, answer is 0.98.

Common Data for Questions 17 and 18:

At a location where the atmospheric pressure is 98 kPa and the ambient temperature is 30°C, the humidity ratio is 0.01 kg/kg of dry air. A high pressure front moves over the location which changes only the atmospheric pressure to 102 *kPa*, while the humidity ratio remains same.

Q17. What is the partial pressure of water vapour in kPa to the first decimal place accuracy before the high pressure front moves in? ________

Solution:

Given:

Atmospheric pressure, $P_{atm} = 98$ kPa,

Ambient air temperature, $T = 30°C$

Humidity ratio (ω) = 0.01 kg/kg of dry air

$$\omega = 0.62198\frac{P_w}{P_a}$$

where, P_w & P_a are the pressure of water vapor and dry air respectively.

$98\times)\ 0.01\text{-}0.01P_w = 0.62198P_w\ (P_{atm} = P_w + P_a)$

$$P_w = \frac{0.98}{0.631988}\text{kPa} = 1.55 \text{ kPa}$$

Hence, the answer is 1.55.

Q18. What is the relative humidity of air under the influence of high pressure front to integer precision in %? _____

Solution:

When the high pressure font moves in, humidity ratio remains the same,

But, $P_{atm} = 102$ kPa.

$$\omega = 0.62198\frac{P_w}{P_a}$$

$$0.01 = 0.62198\times\frac{P_w}{P_a}$$

$$\frac{P_w}{P_a} = 0.016 \quad \text{(i)}$$

Also, $P_{atm} = P_w + P_a$

$$102 = P_w + P_a \quad \text{(ii)}$$

From, eq (i) & eq (ii) we get

$P_w = 1.6$ kPa.

From the Teten's equation

$$P_s = 0.61078\exp\left(\frac{17.27T}{T+273.3}\right) = 0.61078\exp\left(\frac{17.27\times 30}{30+237.3}\right)\text{kPa} = 4.242\text{ kPa}$$

Relative humidity $= \frac{P_w}{P_s}$

$= 1.6/4.242$

$= 0.3771$

$= 37.71\ \%$

Hence, the answer is 37.71.

Common Data for Questions 19 and 20:

A rigid insulated cylinder is divided into two chambers A and B by a thin rigid insulating barrier as shown in the figure below.

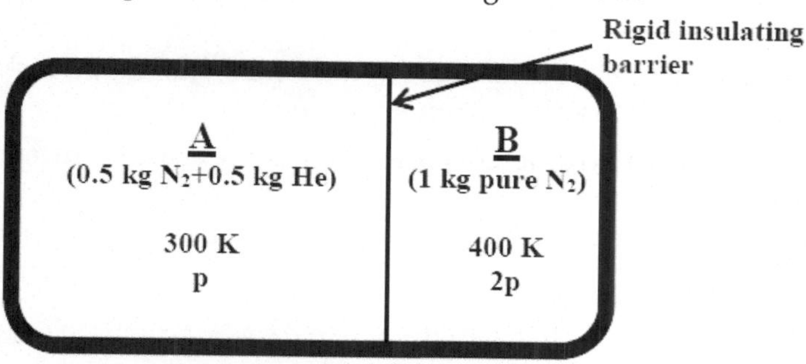

Specific Heat Data:
For He
c_p= 5181 J/kg.K
c_v= 3102 J/kg.K
For N_2
c_p= 1039 J/kg.K
c_v= 742 J/kg.K

Initially, chamber A contains a mixture of 0.5 kg nitrogen and 0.5 kg helium at 300 K while chamber B contains 1 kg of pure nitrogen at 400 K. The pressure in chamber B is twice that in chamber A. The gases and gas mixtures are assumed to be ideal.

Q19. What is the ratio of the volumes of chambers A and B, i.e. V_A/V_B, to first decimal place accuracy? ________

Solution:

Given:

A	B
$0.5\text{kg}N_2 + 0.5\text{kg He}$	$1\text{kg }N_2$
300K	400K
P	2P

P_A (Pressure in chamber A) = P

P_B (Pressure in chamber B) = $2P$

As the cylinder is insulated, dQ = 0

Using ideal gas law, $PV = nRT$

The ratio of volume in chamber A & B, V_A/V_B is:

$$\frac{V_A}{V_B} = \frac{P_B \times n_A T_a R}{P_A \times n_B T_B R}$$

$$= \frac{2 \times \left({}^{0.5}\!/_{28} + {}^{0.5}\!/_{4} \right) \times 300}{1 \times {}^{1}\!/_{28} \times 400} = 6$$

Hence, answer is 6.

Q20. If the barrier is removed and the gases are allowed to mix and reach thermodynamic equilibrium, what is the final temperature of the mixture, to the nearest K? ________

Solution:

If the barrier is removed and the gases are allowed to mix, till, they reach thermodynamic equillibrium, such that they attain a final temperature of t_f.

Also, for chamber 1, comprising of two different gases, the specific heat at constant pressure for the mixture is given by,

$$C_{pmixt} = \frac{m_H Cp_H + m_N Cp_N}{m_H + m_N},$$

= 3110 J/kg.K

As, already stated, system is insulated, i.e. $dQ = 0$,

$$\therefore m_1 C_{pmixt}(T_1 - T_f) = m_2 C_{pN}(T_f - T_2)$$

where m_1, T_1, C_{pmixt} are the mass, initial temperature, specific heat of gases in chamber 1, and m_2, T_2, C_{pN} are the mass, initial temperature & specific heat of gas in chamber 2.

$$T_f = \frac{m_1 C_{pmixt} T_1 + m_2 C_{pN} T_2}{m_1 C_{pmixt} + m_2 C_{pN}}$$

$$T_f = \frac{1 \times 3110 \times 300 + 1 \times 1039 \times 400}{3110 + 1039}$$

= 325.04 K

Hence, the answer is 325.04.

Linked Answer Questions

Statement for Linked Answer Questions 21 and 22:

A combined vapour compression-cum-Brayton cycle is shown below

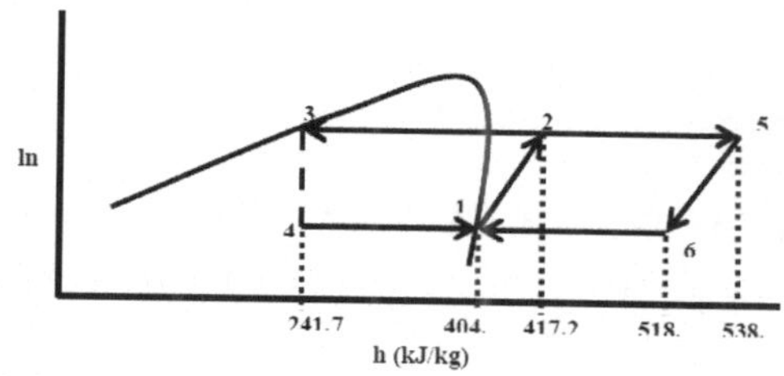

1-2: Isentropic compression, 5-6: isentropic expansion.

The refrigeration system has a cooling capacity of 30 kW and the turbine generates a power of 30 kW.

Q21. What is the mass flow rate of the working fluid through the turbine, in kg/s, to first decimal place accuracy? ________

Solution:

Given:

Cooling capacity of refrigeration system = 30 kW

Power /work generated by turbine, $W = 30$ kW

There is isentropic expansion in turbine.

So, in the graph given, 5-6 represents, turbine.

Enthalpy at the inlet of turbine, $h_5 = 538$ kJ/kg

Enthalpy at the exit of turbine, $h_6 = 518$ kJ/kg

From the SFEE, equation for turbine,

$$\dot{m}h_5 = \dot{m}h_6 + W$$

$$\dot{m}_{wf} = \frac{h_5 - h_6}{W}$$

$$= \frac{30}{(538 - 518)} \text{kg/s}$$

1.5 kg/s

Hence, the answer is 1.5.

Q22. What is the power required to drive the compressor, to the nearest kW? ________

Solution:

The refrigerating effect, $q_0 = h_1 - h_4 = (404 - 241.7)\frac{\text{kJ}}{\text{kg}} = 162.3 \text{ kJ/kg}$

Hence the mass flow rate of the refrigerant,

$$\dot{m}_R = \frac{\text{Cooling capacity } (Q_0)}{\text{Refrigerating effect } (q_0)} = \frac{30}{162.0}\text{kg/s} = 0.1848 \text{ kg/s}$$

Hence the power required to drive the compressor

$$\dot{W}_c = (\dot{m}_R + \dot{m}_{wf})(h_2 - h_1) = (0.1848 + 1.5) \times (417.2 - 404)\text{kW} = 22.24 \text{ kW}$$

Hence, the answer is 22.24.

GATE 2012
THERMODYNAMICS

Q. 1 – Q. 9 carry one mark each.

Q1. Consider a piston-cylinder arrangement containing a gas. This system is heated by placing it on the top of a burner. The system undergoes

(A) a constant volume process

(B) a constant pressure process

(C) an adiabatic process

(D) an isothermal process

Solution:

The answer is (B).

Q2. For a pure substance, at the triple point

(A) only solid and liquid phases co-exist in equilibrium

(B) only liquid and vapour phases co-exist in equilibrium

(C) only solid and vapour phases co-exist in equilibrium

(D) solid, liquid and vapour phases co-exist in equilibrium

Solution:

Triple point is the point of intersection of sublimation line, vaporization line and fusion line on Pressure vs Temperature (P-T) phase diagram where all three phases, i.e. solid, liquid, vapor coexist in equilibrium.

Hence, the answer is (D).

Q3. In a saturated liquid-vapour mixture, the property quality, x is defined as

(A) $x = \dfrac{m_{vapour}}{m_{liquid} + m_{vapour}}$ (B) $x = \dfrac{m_{vapour}}{m_{liquid}}$

(C) $x = \dfrac{m_{liquid}}{m_{liquid} + m_{vapour}}$ (D) $x = \dfrac{m_{liquid}}{m_{vapour}}$

Solution:

When a substance exists as part liquid and part vapor at saturation conditions, its quality (x) is defined as the ratio of the mass of the vapor to the total mass of both vapor and liquid. Quality is zero for the saturated liquid and one for the saturated vapor ($0 \le x \le 1$).

Hence, the answer is (A).

Q4. The slope of a Mollier diagram at constant pressure indicates

(A) enthalpy (B) entropy

(C) internal energy (D) temperature

Solution:

An enthalpy–entropy chart, also known as the h-s chart or Mollier diagram, plots the total heat against entropy.

From the first and second law of thermodynamics following property relation are obtained:

$Tds = dh - vdp$.

Therefore, $\left(\frac{\partial h}{\partial s}\right)_p = T$

The slope of an isobar on the *h-s* coordinates is equal to the absolute temperature.

Hence, the answer is (D).

Q5. If Q_L represents the magnitude of heat transfer from a low temperature reservoir to a cyclic device and Q_H represents the magnitude of heat transfer from a cyclic device to a high temperature reservoir, then for the same Q_L and Q_H, the coefficient performance of a refrigerator (COP_R) and the coefficient performance of a heat pump (COP_{HP}) can be related as

(A) $COP_R = 1 - COP_{HP}$ (B) $COP_{HP} = COP_R + 1$

(C) $COP_R . COP_{HP} = 1$ (D) $COP_{HP} = COP_R - 1$

Solution:

COP = Desired effect / Work input

$COP_R = \frac{Q_L}{W}$

$\therefore Q_H - Q_L = W$

$\therefore COP_{HP} = \frac{Q_H}{W} = \frac{W + Q_L}{W}$

$= 1 + \frac{Q_L}{W} = 1 + COP_R$

Hence, the answer is (B).

Q6. Clausius inequality is written as

(A) $\oint \delta Q > 0$ (B) $\oint \delta Q < 0$

(B) $\oint \frac{\delta Q}{T} \leq 0$ (D) $\oint \frac{\delta Q}{T} > 0$

Solution:

Clausius inequality suggests us the criterion for reversibility and irreversibility of a thermodynamic cycle.

For any cycle, $\frac{\delta Q}{T} \geq ds.$

It can be proven that for a reversible cycle

$\oint \frac{\delta Q}{T} = 0$ (as entropy is a property so its cyclic integral is zero)

Therefore, for a reversible cycle, $\oint \frac{\delta Q}{T} = 0$

And for an irreversible cycle, $\oint \frac{\delta Q}{T} = 0$

Hence, the answer is (C).

Q7. In a Diesel cycle, the ratio of cylinder volumes after and before combustion process is called

(A) cut-off ratio (B) back work ratio

(C) pressure ratio (D) compression ratio

Solution:

Diesel cycle (CI) engines), diesel is injected into the high-pressure combustion chamber as the cylinder reaches TDC and continues after it for a finite duration of time, at a theoretically constant pressure. In this duration, the piston starts moving from the TDC to BDC and this causes a change in volume. Cut-off ratio is defined as the ratio of combustion chamber volume at the end of fuel injection/combustion and the combustion chamber volume at the beginning of fuel injection/combustion.

Hence, the answer is (A).

Q8. The exergy (or availability) of a system at a specified state

(A) depends on the conditions of the system alone

(B) depends on the conditions of the environment alone

(C) depends on the conditions of both the system and environment

(D) depends neither on the conditions of the system nor the environment

Solution:

The (exergy or availability) of a system is the maximum useful work possible during a process that brings the system into equilibrium with its surrounding.

Exergy is a property, which depends on the state of the environment as well as the state of the system.

Hence, the answer is (C).

Q9. In each of the following choices, there are two expressions given. Select the choice that gives, first, the defining expression of volume expansivity and second, the expression of volume expansivity for ideal gases

(A) $\frac{1}{v}\left(\frac{\partial v}{\partial T}\right)_P, \frac{1}{T}$ (B) $-\frac{1}{v}\left(\frac{\partial v}{\partial P}\right)_T, \frac{1}{P}$

(C) $-\frac{1}{v}\left(\frac{\partial v}{\partial P}\right)_T, \frac{1}{P}$ (D) $\frac{1}{v}\left(\frac{\partial v}{\partial T}\right)_P, \frac{1}{P}$

Solution:

Volume expansivity $(\beta) = +\frac{1}{v} \times \left(\frac{\partial v}{\partial T}\right)_P$

For ideal gases,

$$Pv = RT$$

$$v = \frac{RT}{P}$$

$$\left(\frac{\partial v}{\partial T}\right)_P = \frac{R}{P}$$

$$(\beta)_{\text{ideal gas}} = \frac{1}{v} \times \frac{R}{P} = \frac{1}{T}$$

Hence, the answer is (A).

Q. 10 - Q. 22 carry two marks each.

Q10. A certain mass of gas at 0°C is expanded to 81 times its original volume under adiabatic conditions. If ratio of specific heats of the gas, $\gamma = 1.25$, the final temperature of the gas is

(A) – 235°C (B) – 182°C (C) – 91°C (D) 0°C

Solution:

$T_i = 0°C$

$V_2 = 81V_1$

$\Delta Q = 0$

$$\frac{C_P}{C_v} = \gamma = 1.25$$

For, Reversible Adiabatic expansion,

$$\frac{T_2}{T_1}=\left(\frac{V_1}{V_2}\right)^{\gamma-1}$$

$$\frac{T_2}{T_1}=\left(\frac{1}{81}\right)^{0.25}$$

$T_1 = 91$ K

$T_2 = -182°C$

Hence, the answer is (B).

Q11. A sample of gas with initial average kinetic energy, E is heated from 27°C to 327°C. The average kinetic energy after heating is

(A) E (B) 2E

(C) 27E (D) 327E

Solution:

$$K.E=\frac{3}{2}\times k\times T,\ (k = \text{Boltzman's constant})$$

$$E=\frac{3}{2}\times k\times(27+273)$$

$$E=\frac{3}{2}\times k\times 300$$

$E = 450$ k

After heating,

$$E_2=\frac{3}{2}\times 4\times(327\times 273)$$

$$E_2=\frac{3}{2}\times 600$$

$E_2 = 900$ k

After heating, K.E = 2E

Hence, the answer is (B).

Q12. Helium in a piston/cylinder assembly at 20°C and 100 kPa is brought to 400 K in a reversible polytropic process with exponent n = 1.25. Assume helium to be an ideal gas. The molecular mass of helium is 4.003 kg/kmol. The specific work in the process is approximately

(A) – 800 kJ/kg (B) – 788 kJ/kg (C) 788 kJ/kg (D) – 888 kJ/kg

Solution:

The work done for polytropic process,

$$W = \frac{P_1 v_1 - P_2 v_2}{n-1}$$

$$= \frac{RT_1 - RT_2}{n-1}$$

$$= \frac{R(293-400)}{0.25}$$

$$= \frac{8.314 \times (-107)}{0.25}$$

= - 3558.392 kJ/kmol

$$\therefore \, {}^{w}\!/_{m} = -\frac{3558.392}{4.003} \frac{\text{kJ}}{\text{kg}} = -\,888\,\text{kJ/kg}$$

Hence, the answer is (D).

Q13. 32 kg of oxygen is mixed with 28 kg of nitrogen at the same temperature. The gases are at the same pressure of 103 kPa before and after mixing. If *R* is the universal gas constant in kJ/kmol.K, the change in entropy of the mixture is

(A) 1.38*R* (B) 0.69*R*

(C) *R* (D) 0.34*R*

Solution:

$m_1 = 32$

$m_2 = 28$

Now, the number of moles of oxygen, $n_1 = \frac{\mathcal{M}_1}{m_1} = \frac{32}{32} = 1$

Now, the number of moles of nitrogen, $n_2 = \frac{\mathcal{M}_2}{m_2} = \frac{28}{28} = 1$

Let the volume of oxygen before mixing be V_1. Therefore, the equation of state gives

$$P_1 V_1 = n_1 RT_1$$

$$V_1 = \frac{P_1}{RT_1} \quad \text{(i)}$$

Let the volume of nitrogen before mixing be V_2. Therefore, the equation of state gives

$$P_2V_2 = n_2 RT_2$$

$$V_2 = \frac{P_2}{RT_2} \quad \text{(ii)}$$

Since $P_1 = P_2$ and $T_1 = T_2$, therefore from eqs. (i) and (ii)

$$V_1 = V_2 \quad \text{(iii)}$$

Hence, the volume after mixing,

$$V_3 = V_1 + V_2 = 2V_1 \quad \text{(iv)}$$

Since the process involved is a free expansion process, hence the internal energy remains constant. Since $T_1 = T_2$, therefore the final temperature after mixing $T_3 = T_1 = T_2$.

Now the entropy-change of oxygen,

$$(\Delta S)_{\text{oxygen}} = C_v \ln\left(\frac{T_3}{T_1}\right) + R\ln\left(\frac{V_3}{V_1}\right) = R\ln(2)$$

Similarly, the entropy-change of nitrogen,

$$(\Delta S)_{\text{nitrogen}} = C_v \ln\left(\frac{T_3}{T_2}\right) + R\ln\left(\frac{V_3}{V_2}\right) = R\ln(2)$$

Hence, the total entropy change due to the mixing

$$\Delta S = (\Delta S)_{oxygen} + (\Delta S)_{nitrogen} = 2R\ln(2) = 1.38R$$

Hence, the answer is (A).

Q14. Consider two Carnot heat engines A and B operating in series. Engine A receives heat from a reservoir at 1750 K and rejects heat to another reservoir at temperature T. Engine B receives an amount of energy same as that rejected by Engine A from the reservoir at temperature T. Engine B then rejects heat to another reservoir at 320 K. In both cases, the engines produce some amount of work. If the thermal efficiencies of both the engines are the same, then the temperature T is approximately

(A) 848 K (B) 748 K

(C) 648 K (D) 548 K

Solution:

Heat received by reservoir A = Q_1

Heat rejected by reservoir $A = Q_2$

Heat rejected by reservoir $B = Q_3$

Work done by A and B are the same = W

$$\text{Now, } \eta_1 = \frac{W}{Q_1} \text{ and } \eta_2 = \frac{W}{Q_2}$$

$$\eta_1 = \frac{Q_1 - Q_2}{Q_1} = 1 - \frac{T_2}{T_1} = 1 - \frac{T_2}{1750}$$

$$\eta_2 = \frac{Q_2 - Q_3}{Q_2} = 1 - \frac{320}{T_2} = 1 - \frac{320}{T_2}$$

Since $\eta_1 = \eta_2$, therefore

$$1 - \frac{320}{T_2} = 1 - \frac{T_2}{1750}$$

$$-\frac{320}{T_2} = -\frac{T_2}{1750}$$

$$T_2^2 = 56 \times 10^4$$

$$T_2 = \sqrt{560000}$$

$$T_2 = 748 \text{ K}$$

Hence, the answer is (B).

Q15. Joule-Thomson coefficient for a gas, μ_j obeying the relation *p(v-b)=RT* is

(A) $\mu_j = \frac{C_p}{b}$ (B) $\mu_j = \frac{b}{c_p}$

(C) $\mu_j = -\frac{b}{c_p}$ (D) $\mu_j = -\frac{c_p}{b}$

Solution:

Joule-Thomson coefficient μ_j is defined as-change in temperature with respect to pressure at an isenthalpic (h=constant) process, i.e.,

$$\mu_j = \left(\frac{\partial T}{\partial p}\right)_h = \frac{1}{C_p}\left[T\left(\frac{\partial v}{\partial T}\right)_p - v\right]$$

The equation of state for the gas is,

$$p(v - b) = RT$$

$$v = \frac{RT}{p} + b$$

$$\therefore \mu_j = \left(\frac{\partial T}{\partial p}\right)_h = \frac{1}{C_p}\left[T\frac{R}{p} - v\right] = -\frac{b}{C_p}$$

Hence, the answer is (C).

Q16. The correct expression representing Z to be a thermodynamic property is

(A) $Z = pdv$ (B) $Z = vdp$

(C) $Z = pdv + vdp$ (D) $Z = pdv - vdp$

Solution:

Let Z be the thermodynamic property, which is expressed as

$dZ = Mdx+Ndy$

The required condition:

$$\left(\frac{\partial m}{\partial y}\right)_x = \left(\frac{\partial N}{\partial x}\right)_y$$ ... (i) (since a property should be an exact differential)

It is seen that option (C) fulfills the condition of eq. (i).

Hence, the answer is (C).

Common Data Questions

Common Data for Questions 17 and 18:

The vapour pressure of liquid ammonia (in atmosphere) in the vicinity of the triple point can be expressed as

$$\ln p + \frac{3063}{T} = 15.16$$

where temperature T is expressed in K.

In a similar manner, the vapour pressure of solid ammonia can be expressed as

$$\ln p + \frac{3754}{T} = 18.7$$

Take the molecular mass of ammonia to be 17 kg/kmol.

Q17. The temperature and pressure at the triple point are

(A) 295.2 K, 0.69 atm (B) 295.2 K, 0.59 atm

(C) 195.2 K, 0.69 atm (D) 195.2 K, 0.59 atm

Solution:

At triple point pressure and temperature for solid phase and vapour phases coexist.

$$\ln p + \frac{3063}{T} = 15.16 \tag{i}$$

$$\ln p + \frac{3754}{T} = 18.7 \tag{ii}$$

Solving (i) and (ii), we get T= 195.2 K

Substituting T=195.2 K in any of the eqs. (i) & (ii) we get P = 0.59 atm

Hence, the answer is (D).

Q18. The latent heat of vaporization is

(A) 1298 kJ/kg (B) 1398 kJ/kg
(C) 1498 kJ/kg (D) 1698 kJ/kg

Solution:

Differentiating eq. (i) from question no. 17

$$dp/dT = (3063p)/T^2 \quad \text{(i)}$$

According to Clausius-Clapeyron equation,

$dP/dT = h/Tv$, h=latent heat of vaporization (ii)

From eqs. (i) and (ii),

$h/Tv = 3063p/T^2$

h=3063R (iii)

Now R=R_0/ M = 8.314/17 KJ/(kg.K)

Therefore, from eq. (iii)

h=1498 KJ/kg

hence, the answer is (C).

Common Data for Questions 19 and 20:

Consider an ideal reheat cycle utilizing steam. Steam leaves the boiler and enters the turbine at 3 MPa, 400°C (state 3) and then expands to 0.8 MPa (state 4). It is then reheated at constant pressure 0.8 MPa to 400 °C (state 5) and expands to 10 kPa in the low pressure turbine (state 6). The entry to the pump corresponds to saturated liquid state (state 1), and state 2 represents inlet to the boiler. The following data are given:

$h_1 = 191.8$ kJ/kg, $h_3 = 3230.82$ kJ/kg, $h_4 = 2891.6$ kJ/kg, $h_5 = 3267.97$ kJ/kg,

$h_g|_{10\text{ kPa}} = 2584.63$ kJ/kg, $x_6 = 0.92285$, $v_f|_{10\text{ kPa}} = 0.00101$ m^3/kg

Q19. The heat transfer in the boiler is approximately

(A) 4411 kJ/kg (B) 3412 kJ/kg
(C) 3230 kJ/kg (D) 2892 kJ/kg

Solution:

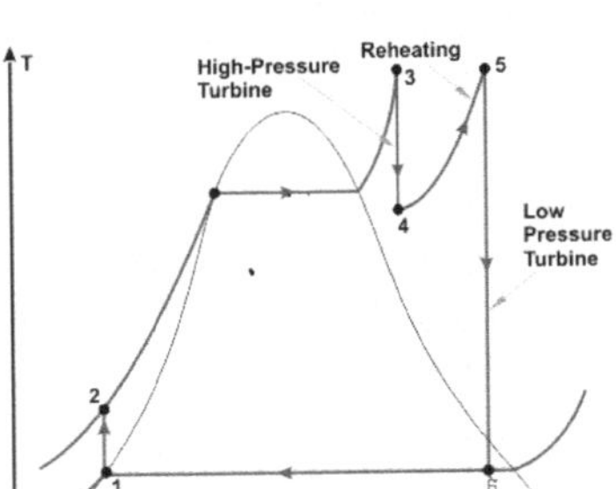

Entry to the H.P Turbine (state 3) h_3=3230.82 kJ/kg

Exit from the H.P. Turbine (state 4) h_4=2891.6 kJ/kg

Entry to the L.P. Turbine (state 5) h_5=3267.97 kJ/kg

Exit from the LP. Turbine (state 6) h_6

Work done by the pump = $v_f(P_2 - P_1)$

$h_2 - h_1 = 0.00101(3\times10^3 - 10)$ kJ/kg

$h_2 - 191.81 = 3.0199$

$h_2 = 194.8299$ kJ/kg

Heat transferred in the boiler = $(h_3 - h_2) + (h_5 - h_4)$

= (3230.82 – 194.8299) + (3267.97 – 2891.6)

= 3412.3601 kJ/kg.

Hence, the answer is (B).

Q20. The net work done in the cycle is approximately

(A) 1004 kJ/kg (B) 1104 kJ/kg

(C) 1204 kJ/kg (D) 2004 kJ/kg

Solution:

Net work done during the cycle = (work done by (Turbine –Pump))

Work done by Turbine, $w_T = (h_3 - h_4) + (h_5 - h_6)$ (i)

$h_6 = h_f + x.h_{fg}$

= 191.81 + 0.92285(2584.63 - 191.81)

= 2400.023 kJ/kg

From eq. (i)

$w_T = (h_3 - h_4) + (h_5 - h_6)$

= {(3230.82- 2891.6) + (3267.97-2400.023)} kJ/kg

= 1207.167 kJ/kg

Work done by Pump, $w_P = h_2 - h_1$ = (194.8299-191.81) kJ/kg = 3.0199 kJ/kg

Net work done during the cycle= $w_T - w_P$ =(1207.167-3.0199) kJ/kg =1204.1471 kJ/kg.

Hence, the answer is (C).

Linked Answer Questions

Statement for Linked Answer Questions 21 and 22:

A piston-cylinder arrangement as shown in the figure initially contains air at 150 kPa and 400°C. The arrangement is allowed to cool to the ambient temperature of 20°C. The characteristic gas constant for air is 0.287 kJ/kg.K. The cylinder wall has stops of negligible thickness that can prevent the piston from moving down. The stops are 1 m from the inner side of the base surface of the cylinder. At the initial state, the piston is resting 1 m above the stops.

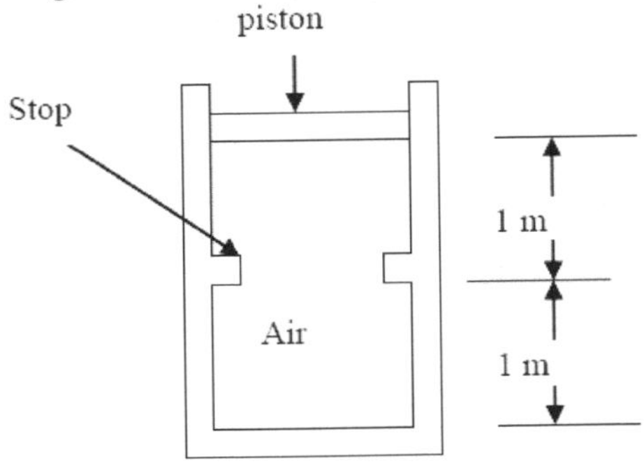

Q21. The final pressure in the cylinder is

(A) 130.7 kPa (B) 150 kPa

(C) 200.7 kPa (D) 230.7 kPa

Solution:

$$\frac{P_1 V_1}{T_1} = \frac{P_2 v_2}{T_2}$$

$$V_2 = \frac{V_1}{2}$$

$$P = \frac{150 \times 2 \times 293}{673} \text{ kPa}$$

$P_2 = 130.6$ kPa

Hence, the answer is (A).

Q22. The specific work done by the air during the process is

(A) –26.67 kJ/kg (B) 26.67 kJ/kg

(C) 49.5 kJ/kg (D) –96.67 kJ/kg

Solution:

Let the polytropic index of the compression process be n. Hence

$$P_1 V_1^n = P_2 V_2^n$$

$$n = \frac{\ln\left(\frac{P_2}{P_1}\right)}{\ln\left(\frac{V_1}{V_2}\right)} = (-0.19)$$

The specific work is given by

$$w = \frac{R(T_1 - T_2)}{n-1} = \frac{0.287 \times 380}{-0.19 - 1} \text{ kJ/kg} = -91.64 \text{ kJ/kg}$$

Hence, the answer is (D).

GATE 2011
THERMODYNAMICS

Note: Usual notations have been used for Thermodynamic variables.

Useful Data:

Universal gas constant, R_0 =8.314 kJ/kmol K; Acceleration due to gravity, g= 9.81 ms^{-2},

Molecular mass of air, Mair = 29 kg/kmol; Specific heat at constant pressure of air, c_p=1.008 kJ/kgK,

Ratio of Specific Heat for air γ = 1.4, Assume air to be a perfect gas unless specified otherwise.

Q.1 – Q.9 carry one mark each

Q1. Identify the group containing the appropriate match of items in List – I and List – II

	LIST - I		LIST -II
K	A jet engine in flight	**P**	Closed System
L	Water being heated in a sealed container	**Q**	Control Volume
M	Internal Energy	**R**	Intensive Property
N	Specific Entropy	**S**	Extensive Property

(A) K-P; L-Q; M-R; N-S
(B) K-Q; L-P; M-R; N-S
(C) K-Q; L-P; M-S; N-R
(D) K-P; L-Q; M-S; N-R

Solution:

The answer is (C).

Q2. The mass, internal energy, pressure, and volume of a system are 10 kg, 100 kJ, 1 bar, and 1 m^3, respectively. The value of specific enthalpy, in kJ/kg of the system is

(A) 10 (B) 20 (C) 30 (D) 40

Solution:

Specific enthalpy, $h = \dfrac{U+PV}{m} = \dfrac{100+100\times 1}{10}\ \text{kJ/kg} = 20\ \text{kJ/kg}$

Hence, the answer is (B).

Q3. A heat pump extracts heat at the rate of 100 kW from a low temperature reservoir and delivers heat at the rate of 160 kW to a high temperature reservoir. The COP of the heat pump is

(A) 1.60 (B) 1.67 (C) 2.60 (D) 2.67

Solution:

$$\text{COP}_{\text{HP}} = \frac{Q_k}{W} = \frac{160}{160-100} = 2.67$$

Hence, the answer is (D).

Q4. At a certain pressure, the specific volumes (m³/kg) of saturated liquid, saturated vapour and wet steam are 1.1565×10^{-3}, 0.1274, and 0.1 respectively. The quality of wet steam is

(A) 0.805 (B) 0.783 (C) 0.776 (D) 0.673

Solution:

$$x = \frac{v - v_f}{v_g - v_f} = \frac{0.1 - 0.0011567}{0.1274 - 0.0011564} = 0.783$$

Hence, the answer is (B).

Q5. Which one of the following expressions represents the Joule – Thompson coefficient?

(A) $\left(\frac{\partial T}{\partial p}\right)_h$

(B) $\left(\frac{\partial T}{\partial v}\right)_h$

(C) $\left(\frac{\partial p}{\partial h}\right)_s$

(D) $\left(\frac{\partial s}{\partial T}\right)_p$

Solution:

Joule-Thomson coefficient is defined as, $\mu_f = \left(\frac{\partial T}{\partial p}\right)_i$

Hence, the answer is (A).

Q6. Specific heat at constant pressure (C_p) of Helium is 5.19kJ/K Kg and its molecular mass is 4 kg/kmol. The specific heat at constant volume in kJ/K kg is

(A) 1.11 (B) 2.11

(C) 3.11 (D) 4.11

Solution:

$C_P - C_V = R$

$C_V = C_P - R$

$= 5.19-(8.314/4)$

$= 3.11$

Hence, the answer is (C).

Q7. An evacuated, rigid adiabatic tank is slowly filled with air from a supply line supplying air at constant pressure, p_L and Temperature T_L. The temperature of air in the tank at the end of the filling process will be

(A) Greater than T_L

(B) Equal to T_L

(C) Less than T_L

(D) The Average of ambient temperature and T_L

Solution:

The unsteady energy equation applied to a control volume is,

$$\frac{dE_V}{dt} = \frac{d}{dt}\left(U + \frac{1}{2}mV^2 + mgz\right)_{CV}$$

$$\frac{d}{dt}\left(U + \frac{1}{2}mV^2 + mgz\right)_{CV} = \frac{dm_1}{dt}\left(h_1 + \frac{V_1^2}{2} + gZ_1\right) - \frac{dm_2}{dt}\left(h_2 + \frac{V_2^2}{2} + gZ_2\right) + \frac{\delta Q}{dt} - \frac{\delta W}{dt} \quad \text{(i)}$$

For the present problem $\frac{dm_2}{dt} = \frac{\delta Q}{dt} = \frac{\delta W}{dt} = \left(\frac{1}{2}mV^2\right)_{CV} = (mgz)_{CV} = 0$ (ii)

Also, $U = C_v T, h_1 = C_p T_L$, (iii)

Therefore, integrating eq. (i) from the initial time to the final time

$$C_v \Delta T = \Delta m\left(C_p T_L + \frac{V_1^2}{2} + gZ_1\right) \quad \text{(iv)}$$

Since the tank is being filled up, therefore $\Delta m > 0$. Hence from eq. (iv)

$\Delta T > 0$

$T_f - T_L > 0$

$T_f > T_L$,

where T_f is the final temperature in the tank.

Hence, the answer is (A).

Q8. An ideal gas undergoes a cyclic process as shown in p–V diagram below:

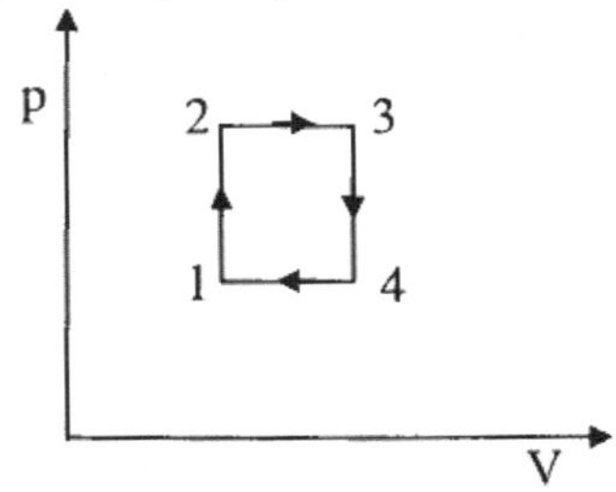

The same cycle, represents in T-S diagram is

(A)

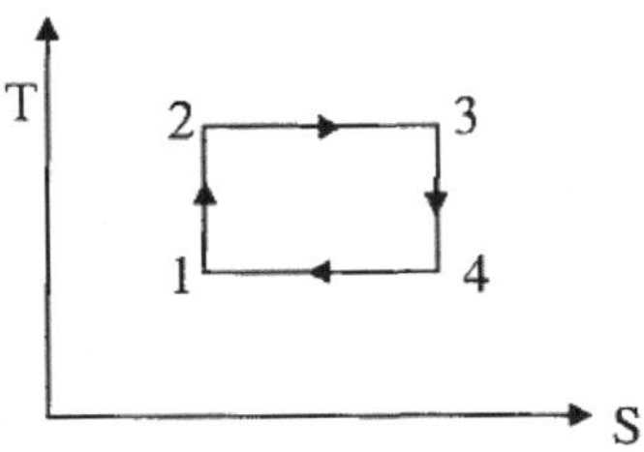

(B)

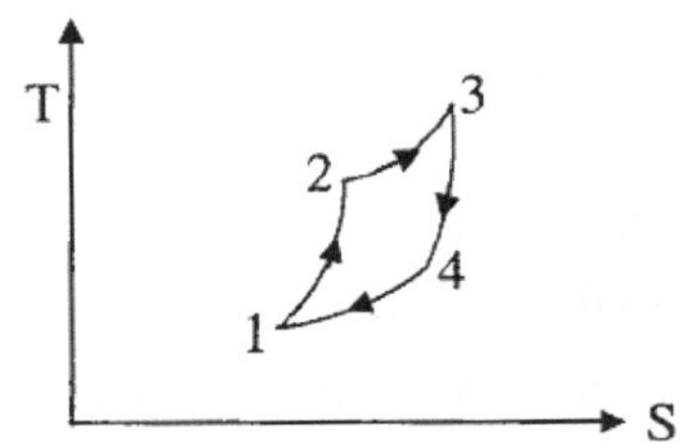

(C)

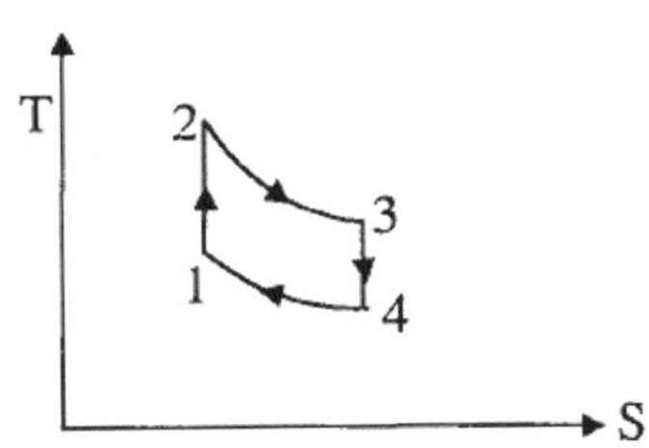

(D)

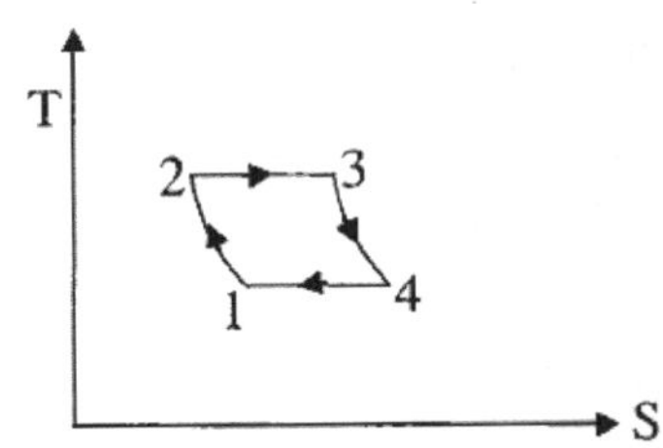

Solution:

For an ideal gas,

$$\left(\frac{\partial T}{\partial s}\right)_p = \frac{T}{C_p} \tag{i}$$

And,

$$\left(\frac{\partial T}{\partial s}\right)_v = \frac{T}{C_v} \tag{ii}$$

Since $C_p > C_v$, hence from eqs. (i) and (ii)

$$\left(\frac{\partial T}{\partial s}\right)_v > \left(\frac{\partial T}{\partial s}\right)_p \tag{iii}$$

Hence, the answer is (B).

Q9. Consider 3 identical tanks A, B and C, shown below. What is the pressure p in tank C?

Tank A	Tank B	Tank C
1 m^3 x kg nitrogen 1 bar, 300 K	1 m^3 y kg oxygen 1 bar, 300 K	1 m^3 x kg nitrogen + y kg oxygen 300 K, p = ?

(A) 1 bar (B) 1.5 bar (C) 2 bar (D) 2.5 bar

Solution:

Tank A

$P_A V_A = n_A R T_A$

$1 \times 1 = n_A \times R \times 300$

$n_A = 1/(300R)$

Tank B

$P_B V_B = n_B R T_B$

$1X1 = n_B \times R \times 300$

$n_B = 1/(300R)$

Tank C

$P_C V_C = n_C R T_C$

$1 X P_C = (n_A + n_B) \times R \times 300$

$= \{1/(300R) + 1/(300R)\} \times R \times 300$

$= 2/(R \times 300) \times R \times 300$

$P_C = 2$ bar

Hence, the answer is (C).

Q.10 – Q.22 carry 2 marks each

Q10. The thermodynamic property, x, of a thermometer varies with temperature, t, according to the relation $t = ax^2 + b$, where t is in (°C), x is in cm, a and b are constants. At ice point (0°C) and steam point (100 °C), the value of x are 5 cm and 20 cm, respectively. When the thermometer is brought in contact with the heated body, the value of x is recorded as 15 cm. The temperature of the heated body in is

(A) 83.3 (B) 73.3 (C) 63.3 (D) 53.3

Solution:

$$t = a \times x^2 + b$$

$$0 = a \times 5^2 + b \qquad \text{(i)}$$

$$100 = a \times 20^2 + b \tag{ii}$$

Solving Equation (i) and (ii)

$a = 0.267$ $b = -6.67$

$t = 0.267 \times x^2 - 6.67$

At x =15 cm, the temperature t = 53.3°C

Hence, the answer is (D).

Q11. Air at 10 bar and 400 K is flowing with a velocity of 20 m/s through a pipe of uniform cross sectional area of 20 cm². The mass flow rate in the pipe is approximately

(A) 0.25 kg/s (B) 0.35 kg/s (C) 2.5 kg/s (D) 3.5 kg/s

Solution:

Mass flow rate can also be calculated by:

$$\dot{m} = \rho \times V = \rho \times V \times A = \frac{P\mathcal{M}}{RT} \times V \times A$$

ρ = mass *density* of the fluid, kg/m³

V = *Volume flow rate*, m³/s

V = *Flow velocity* of the mass elements, m/s

A = *cross-sectional vector area*/surface, m²

P = Pressure, M = Molecular mass and T = Temperature

R= 8.314 J/K mol = 8.314 x 10^3 L Pa/K mol = 8.314 x 10^{-2} L bar/K mol

$$\dot{m} = \frac{P\mathcal{M}}{RT} \times V \times A = \frac{10 \times 29}{8.314 \times 10^{-2} \times 400} \times 20 \times 20 \times 10^{-4} \text{ kg/s} = 0.35 \text{ kg/s}$$

Hence, the answer is (B).

Q12. The condition of steam at inlet and exit of a perfectly insulated steam turbine running under steady – state conditions as follows

At inlet, specific enthalpy = 3230 kJ/kg, velocity =160 m/s

At exit, specific enthalpy = 2660 kJ/kg, velocity =100 m/s

Neglecting changes in the Potential Energy, the work done by the steam turbine is approximately

(A) 570kJ/kg (B) 578kJ/kg
(C) 630kJ/kg (D) 638kJ/kg

Solution:

Change in kinetic energy: $\Delta KE = \frac{V_{in}^2}{2} - \frac{V_{out}^2}{2} = \frac{160^2}{2} - \frac{100^2}{2} = 7.8 \text{ kJ/kg}$

Work done is obtained by applying the steady flow energy equation:

$$h_{in} + \frac{V_{in}^2}{2} + gZ_{in} + q = h_{out} + \frac{V_{out}^2}{2} + gZ_{out} + w$$

Neglecting the change in potential energy and considering that the turbine is perfectly insulated

$$w = h_{in} - h_{out} + \frac{V_{in}^2}{2} - \frac{V_{out}^2}{2}$$

=(3230-2660+7.8) kJ/kg

=577.8 kJ/kg

Hence, the answer is (B).

Q13. The thermal efficiency of a reversible heat engine operating between two thermal reservoirs is η_{max}. The coefficient of performance of a reversible refrigeration cycle operating between the same two reservoirs is

(A) $\left(\frac{1}{\eta_{max}}\right) - 1$ (B) $\left(\frac{1}{\eta_{max}}\right) + 1$

(C) $\left(\frac{1}{\eta_{max}}\right)$ (D) $1 - \left(\frac{1}{\eta_{max}}\right)$

Solution:

$$\eta = \frac{Q_H - Q_L}{Q_H}, COP_R = \frac{Q_L}{Q_H - Q_L}$$

$$COP_R + 1 = \frac{Q_L}{Q_H - Q_L} + 1 = \frac{Q_L + Q_H - Q_L}{Q_H - Q_L} = \frac{Q_H}{Q_H - Q_L} = \frac{1}{\eta}$$

Hence, $COP_R = \frac{1}{\eta} - 1$

Hence, the answer is (A).

Q14. Given for water,

At 190 °C, saturation pressure = 12.54 bar

At 210 °C, saturation pressure = 19.06 bar

At 200 °C, specific volume of saturated liquid = 1.1565×10^{-3} m³/kg

At 200 °C, specific volume of saturated vapour = 0.1274 m³/kg

Using Clapeyron equation, the enthalpy of vaporization of water at 200°C is approximately

(A) 1857 (B) 1924 (C) 1947 (D) 2025

Solution:

$$\frac{dp}{dT} = \frac{h_{fg}}{T\left(v_g - v_f\right)}$$

$$\frac{(19.06 - 12.54) \times 10^5}{(483 - 463)} = \frac{h_{fg}}{473 \times (0.1279 - 1.1565 \times 10^{-3})}$$

$h_{fg} = 1947$ kJ/kg

Hence, the answer is (C).

Q15. An ideal vapour compression refrigeration cycle uses R-134a as working fluid. The condenser pressure is 5 bar and evaporator pressure is 60 kPa. The refrigerant enters the compressor as saturated vapour and its specific enthalpy at the compressor exit is 267.98 kJ/kg. The other relevant data regarding R-134a are given below

Pressure, bar	**h_f, kJ/kg**	**h_g, kJ/kg**	**s_f, kJ/kg K**	**S_g, kJ/kg K**
0.6	3.46	224.72	0.0147	0.9520
5	71.33	256.07	0.2723	0.9117

The COP of the cycle is approximately

(A) 2.55 (B) 3.55 (C) 4.11 (D) 5.11

Solution:

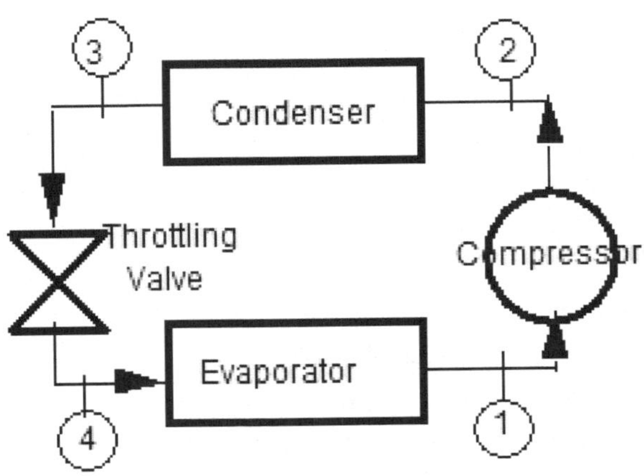

$$COP = \frac{Q_L}{W_{net,in}} = \frac{Q_{evap}}{Q_{comp}} = \frac{h_1 - h_4}{h_2 - h_2} = \frac{224.72 - 71.33}{267.98 - 224.72} = 3.55$$

Hence, the answer is (B).

Q16. One kilomole of hydrogen (M = 2kg/mole) is mixed with certain number of kilomoles of argon (M = 40kg/mole) such that the mass fraction of argon in the resultant mixture is 0.8. The number of kilomoles of argon in the mixture is

(A) 0.05 (B) 0.10

(C) 0.15 (D) 0.20

Solution:

$$m_{Total} = m_A + m_B = m_{H_2} + m_{Ar}$$

Let number of kilo moles of argon in mixture = x

The mass fraction of each component becomes

$$m_{H_2} = N_{H_2} M_{H_2}$$

$$m_{Ar} = N_{Ar} M_{Ar}$$

$$mf_{Ar} = \frac{m_{Ar}}{m_{Total}} = \frac{m_{Ar}}{m_{H_2} + m_{Ar}} = \frac{N_{Ar} M_{Ar}}{N_{H_2} M_{H_2} + N_{Ar} M_{Ar}} = \frac{40x}{2 + 40x} = 0.8$$

Hence number of kilo moles of argon in mixture is x= 0.2

Hence, the answer is (D).

COMMON DATA QUESTION

Common Data questions 17 and 18:

A steam power plant operates on an ideal Rankine cycle between pressures of 5 MPa (Boiler) and 10 kPa (condenser). Water enters the pump at a specific enthalpy of 191.83 kJ/kg, the specific volume is 0.00101 m^3/kg and the steam enters the turbine at an enthalpy of 4612.2 kJ/kg, and leaves the turbine as saturated vapour at 10 kPa. The enthalpy of vaporization of water at 10 KPa is 2392.8 kJ/kg.

Q17. The net work done per cycle (kJ/kg) is approximately

(A) 2023 (B) 2123

(C) 2223 (D) 2323

Solution:

Work supplied to the pump,

$w_P = v_f (p_2 - p_1) = \{0.00101 \times (5000 - 10)\}$ kJ/kg = 5.04 kJ/kg

Work done by the turbine,

$w_T = h_3 - h_4 = \{4612.2 - (191.83 + 2392.8)\}$ kJ/kg = 2027.57 kJ/kg

Hence, the net work done by the cycle,

$w_{net} = w_T - w_P$ = (2027.57-5.04) kJ/kg = 2022.53 kJ/kg

Hence, the answer is (A).

Q18. The thermal efficiency of the cycle is,

(A) 35.8% (B) 40.8%

(C) 45.8% (D) 50.8%

Solution:

Heat supplied,

$Q = h_3 - h_2 = h_3 - (h_1 + w_P) = \{4612.2 - (191.83 + 5.04)\}$ kJ/kg = 4415.83 kJ/kg

Therefore, the thermal efficiency

$$\eta = \frac{w_{net}}{Q} = \frac{2023}{44415.83} = 0.458$$

Hence, the answer is (C).

Common Data questions 19 and 20:

At the beginning of compression process of an air–standard Otto cycle, the pressure, temperature and volume are 1 bar, 290 K, and 400 cm^3, respectively. The compression ratio is 8. The maximum temperature attained in the cycle is 2200 K.

Q19. The heat transferred (kJ) during the heat addition process is approximately

(A) 531 (B) 53.1

(C) 5.31 (D) 0.531

Solution:

$$T_2 = T_1\left(\frac{V_1}{V_2}\right)^{\gamma-1} = 290\times(8)^{1.4-1}\ \text{K} = 666.25\ \text{K}$$

The characteristic gas constant for air,

$$R_{air} = \frac{R_0}{\mathcal{M}_{air}} = \frac{8.314}{29}\ \text{kJ/kg. K} = 0.287\ \text{kJ/kg. K}$$

The specific heat of air at constant volume,

$$(C_v)_{air} = \frac{R_{air}}{\gamma-1} = \frac{0.287}{1.4-1}\ \text{kJ/kg. K} = 0.7175\ \text{kJ/kg. K}$$

The mass of air,

$$m = \frac{p_1V_1}{RT_1} = \frac{10^2\times400\times10^{-6}}{0.287\times290}\ \text{kg} = 4.8\times10^{-4}\ \text{kg}$$

Therefore, the heat added

$Q = m(C_v)_{air}(T_3 - T_2) = 4.8\times10^{-4}\times0.7175\times(2200 - 666.25)$ kJ = 0.53 kJ

Hence, the answer is (D).

Q20. The thermal efficiency of the cycle is

(A) 41.5% (B) 46.5%

(C) 51.5% (D) 56.5%

Solution:

The thermal efficiency of an Otto cycle is

$$\eta_{Otto} = 1 - \frac{1}{r_c^{\gamma-1}} = 1 - \frac{1}{8^{1.4-1}} = 0.565$$

Hence, the answer is (D).

LINKED ANSWER QUESTIONS

Statement for Linked Answer Questions 21 and 22:

An ideal gas undergoes a cyclic process consisting of the following three processes:

Process1-2: Compression process with pV= constant

Process2-3: Constant Pressure

Process 3-1: Constant volume; U_3-U_1=3549kJ

Changes in kinetic and potential energies are neglected

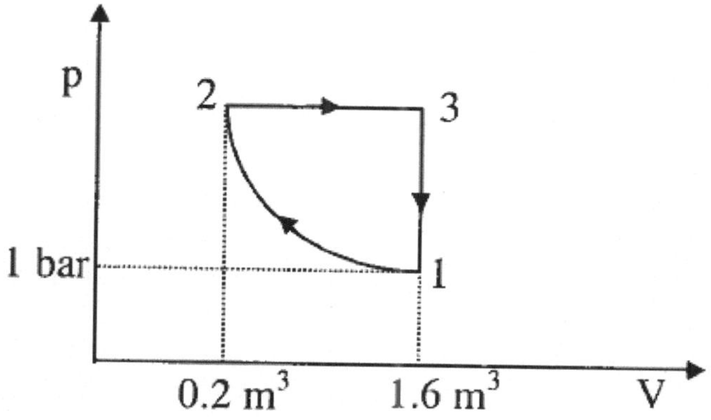

Q21. The work done (kJ) during process 2-3 is

(A) 1120 (B) 1220

(C) 1320 (D) 1420

Solution:

$$p_2 = p_1\left(\frac{V_1}{V_2}\right) = 1\times\left(\frac{1.6}{0.2}\right)\text{bar} = 8 \text{ bar}$$

Therefore, the work done during the process 2-3 is

$W_{2-3} = p_2 (V_3 - V_2) = 8 \times 10^2 \times (1.6 - 0.2)$ kJ = 1120 kJ

Hence, the answer is (A).

Q22. The heat transferred (KJ) during process 2-3 is

(A) 4269 (B) 4469

(C) 4569 (D) 4669

Solution:

It may be recalled here that , $U_3 - U_2 = U_3 - U_1 = 3549$ kJ, since $U_1 = U_2$.

$\therefore Q_{2-3} = U_3 - U_2 + W_{2-3}$

$= (3549 + 1120)$ kJ

$= 4669$ kJ

Hence, the answer is (D).

GATE 2010
THERMODYNAMICS

Q.1-Q.9 carry one mark each.

Q1. Match the items in Group I for their correctness with the corresponding appropriate terms given in Groups II and III.

Group I	Group II	Group III
P: Pressure	1: Path dependent quantity	X: Intensive property
Q: Heat	2: Path independent quantity	Y: Extensive property

(A) P, 1, X (B) P, 2, X (C) Q, 1, X (D) Q, 2, Y

Solution:

Pressure is an intensive property and a property is point function and not a path function.

Hence, the answer is (B).

Q2. An object of mass 'm' in a wooden box having mass 'M' falls through a height 'h' under the influence of gravity in vacuum. The work done by the object on the box is

(A) 0 (B) mgh (C) Mgh (D) (m+M)gh

Solution:

Because vacuum does not exert any external force.

Hence, the answer is (A).

Q3. An ideal gas is known to obey following relationships: $u = 200 + 0.718T$ and $Pv = 0.287(T + 273)$, where u is specific internal energy (kJ/kg), T is temperature (°C), P is pressure (kPa) and v is specific volume (m^3/kg). Specific heat (in kJ/kg-K) at constant pressure is

(A) 0.287 (B) 0.431 (C) 0.718 (D) 1.005

Solution:

$h = u + Pv$

$Cp\,(T + 273) = 200 + 0.718T + 0.287(T + 273) = 278.35 + 1.005T$

Let $T = 0°C$

$Cp = 278.35/273$ kJ/kg-K

$= 1.019$ kJ/kg-K

The closest answer among the given options is 1.005 kJ/kg-K

Hence, the answer is (D).

Q4. A heat pump which operates in a cycle, extracts heat energy from the cold reservoir and supplies the same amount of energy to the hot reservoir. Which of the following statement hold for this process?

(A) This process violates both the first and the second law

(B) This process violates the first law but not the second law

(C) This process violates the second law but not the first law

(D) This process does not violate both first and second law

Solution:

The answer is (C).

Q5. An insulated rigid container having 1m³ volume has two compartments having equal volume separated by a thin membrane. Half of the container is filled with helium ($R = 2.08$ kJ/kg-K, $C_p = 5.19$ kJ/kg-K and $C_v = 3.11$ kJ/kg-K), while the remaining half is empty. Suddenly the membrane ruptures and helium fills the whole volume of the container. Temperature and pressure of helium before rupture are 500°C and 0.1 MPa respectively. The change in the entropy of helium is

(A) 0.019kJ/K (B) 0.045 kJ/K

(C) 0.112kJ/K (D) 0.675 kJ/K

Solution:

$$m = \frac{p_1 V_1}{RT_1} = \frac{10^2 \times 0.5}{2.08 \times 773} \text{ kg} = 0.0311 \text{ kg}$$

Since the process is a free-expansion process, hence $\Delta U = 0$, or $T_2 = T_1$.

The change in entropy is given by

$$\Delta S = mR \ln\left(\frac{V_2}{V_1}\right) = 0.0311 \times 2.08 \times \ln(2) \text{ kJ/K} = 0.045 \text{ kJ/K}$$

Hence, the answer is (B).

Q6. 1 kg of methane is enclosed in a cylinder having volume 6.4 litres and is maintained at a temperature of 13°C and pressure of 18.56 MPa. If molecular weight of methane is 16 kg/kmol (for methane, critical pressure = 4.64 MPa, critical temperature is 191.1 K; universal gas constant is 8.314 kJ/kmol-K), compressibility factor, *Z*, is

(A) 0.375 (B) 0.8

(C) 1.25 (D) 2.66

Solution:

$$Z = PV/nRT$$

$$Z = \frac{18.56 \times 10^3 \times 6.4 \times 10^{-3}}{\frac{1}{16} \times 8.314 \times 286} = 0.8$$

Hence, the answer is (B).

Q7. $\left(\frac{\partial P}{\partial T}\right)_s$ **is equal to**

(A) $\left(\frac{\partial S}{\partial V}\right)_P$ (B) $-\left(\frac{\partial S}{\partial V}\right)_P$ (C) $\left(\frac{\partial S}{\partial V}\right)_T$ (D) $-\left(\frac{\partial S}{\partial V}\right)_T$

Solution:

The answer is (A).

Q8. A rigid spherical vessel contains 1 kg of wet steam of quality x at pressure P_1. This is shown by point A on the T - v diagram. Heat is transferred to the vessel to form superheated steam at pressure P_2, and temperature T_2 as shown by point B.

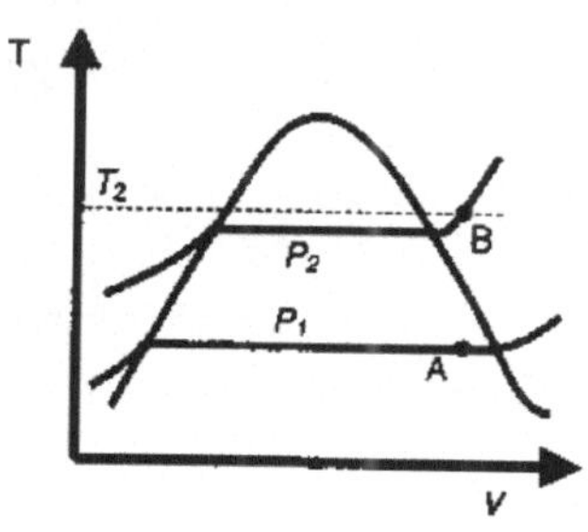

Specific enthalpy and specific internal energy corresponding to the saturated water and saturated vapour at pressures P_1 and P_2 as well as at points A and B are given by

Property	Saturated liquid		Saturated vapour		Point A	Point B
	Pressure P_1	Pressure P_2	Pressure P_1	Pressure P_2		
Specific enthalpy (kJ/kg)	h_{f1}	h_{f2}	h_{g1}	h_{g2}	1_A	h_B
Specific internal energy (kJ/kg)	u_{f1}	u_{f2}	u_{g1}	u_{g2}	u_A	u_B

Heat transferred to the steam is

(A) h_B-h_A (B) h_B-h_{f1} (C) u_B-u_A (D) u_B-u_{f1}

Solution:

The system is a closed thermodynamic system with rigid walls, hence the work done $W = 0$.

Now, the first law of thermodynamics for a closed system gives

$$q = \Delta u + w = u_B - u_A$$

Hence, the answer is (C).

Q9. Determine the correctness or otherwise of the following Assertion [a] and the Reason [r] Assertion: Carnot cycle is not used in vapour power cycles

Reason: Pumping of a two phase mixture is difficult

(A) Both **[a]** and **[r]** are true and **[r]** is a reason for **[a]**

(B) Both **[a]** and **[r]** are true but **[r]** is NOT a reason for **[a]**

(C) **[a]** is true but **[r]** is false

(D) Both **[a]** and **[r]** are false

Solution:

The answer is (A).

Q.10-Q.22 carry two marks each.

Q10. A slab of mass 2000 kg and volume 0.2 m³ is raised slowly in the vertical direction by a massless rope through a height of 2 m from the bottom of a fresh-water lake. Depth of water in the lake is 10 m. If the density of fresh water is 1000 kg/m³ and acceleration due to gravity is 10 m/s², work done by water on the block is

(A) -40 kJ (B) 0 kJ (C) 4 kJ (D) +40 kJ

Solution:

Work = density × gravity × volume × distance = $10^3 \times 10 \times 0.2 \times 2$ J = 4 kJ

Hence, the answer is (C).

Q11. A bulb is connected to a U-tube with long vertical column as shown in the figure below. Cross sectional area of both sides of the U-tube is 3 cm². Water is poured in the right side open column of the U-tube such that the trapped air in the left column and the above water has a volume of 1000 cc and a pressure of 105 kPa. Ambient pressure is 100 kPa. Density of water is 1000 kg/m³. Acceleration due to gravity may be taken as 10 m/s².

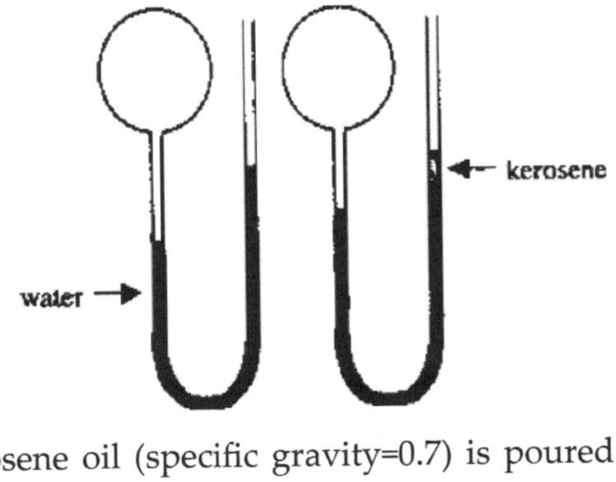

When 90 cc of kerosene oil (specific gravity=0.7) is poured on the right side of

the U-tube, water level in the left column rises by 10 cm. The pressure in the bulb becomes

(A) 105.1 kPa (B) 108.3 kPa (C) 112.4 kPa (D) 119.7 kPa

Solution:

The process can be considered as isothermal. Hence for the air.

$P_1V_1 = P_2V_2$

$$P_2 = P_1\left(\frac{V_1}{V_2}\right) = 105\times\left(\frac{1000}{1000-3\times10}\right)\text{kPa} = 108.3\text{ kPa}$$

Hence, the answer is (B).

Q12.

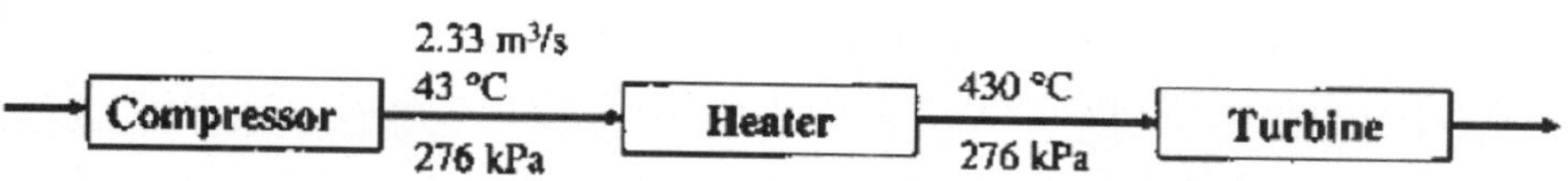

Air (R = 287 J/kg-K, C_p = 1005 J/kg-K and γ = 1.4) flows sequentially through a compressor, a heater and a turbine as shown in the figure. Volume flow rate of air coming out from the compressor is 2.33 m³/s when pressure and temperature are 276 kPa and 43°C respectively. Air is then heated at same pressure to 430°C in a heater, air flows through a turbine which produces 1800 kW of power. Heat loss from turbine to the surrounding is 90 kW. Air temperature at the turbine exit is

(A) 156.4°C (B) 181.6°C

(C) 223.7°C (D) 678.4°C

Solution:

Mass flow rate $\dot{m} = \frac{PQ}{RT}$ = 276 × 1000 × 2.33/{287 × (273 + 43)} = 7.09 kg/s

$C_p\Delta T = Q\text{-}W$

7.09 × 1.005 × (t-430) = -90-1800

t = 164.8°C

The nearest option is 156.4°C.

Hence, the answer is (A).

Q13. Availability per unit mass associated with air (R = 287 J/kg-K, C_p = 1005 J/kg-K and γ = 1.4) flowing from a reservoir at 10 atm and 25°C when atmosphere is at 1 atm and 25°C is (Neglect changes in the potential and the kinetic energies).

(A) 98.4 kJ/kg (B) 196.9 kJ/kg (C) 492.3 kJ/kg (D) 689.14 kJ/kg

Solution:

Work = $RT\ln(P_1/P_2)$ = 287 × ln(10/1) × 298 J/kg = 196.93 kJ/kg

Hence, the answer is (B).

Q14. In an air standard Diesel cycle, compression ratio is 14. At the beginning of compression of air (R=287 J/kg-K, C_p = 1005 J/kg-K and γ = 1.4) the temperature is 27°C and the pressure is 1 bar. If the specific volume after heat addition is two times the specific volume after compression, heat added (in kJ per kg of air) at constant pressure is

(A) 555.5 (B) 622.7 (C) 767.8 (D) 866.4

Solution:

$r_k = V_1/V_2 = 14$

$r_c = V_3/V_2 = 2$

$T_2 = T_1 \times (V_1/V_2)^{(\gamma-1)} = 862.13$ K

$T_3 = T_2 \times V_3/V_2 = 1724.26$ K

Therefore, the heat added per kg of air at constant pressure

$Q = C_p(T_3 - T_2) = 1005 \times (1724.26 - 862.13)$ J/kg = 866.4 kJ/kg

Hence, the answer is (D).

Q15. A mixture of Freon and air is supplied for cleaning a chamber. The mixture contains 70% by volume of air and 30% by volume of Freon. Specific heat ratios for Freon and air are 1.1 and 1.4 respectively. Molecular mass of Freon is 200 g/mole and that of air is 30 g/mole. Temperature of gas is 300 K. If universal gas constant is 8.314 J/mole-K, specific heat ratio of the mixture is

(A) 1.16 (B) 1.21

(C) 1.25 (D) 1.31

Solution:

Component	Percent volume	Mole fraction	Molecular mass	Mass of gas
Freon	30%	0.3	200 g/mole	200×0.3 = 60 g
Air	70%	0.7	30 g/mole	30 = 21 g

$$R_{air} = \frac{R_0}{\mathcal{M}_{air}} = \frac{8.314}{30} \text{ J/g. K} = 0.277 \text{ J/g. K}$$

$$(C_P)_{air} = \frac{\gamma_{air} R_{air}}{\gamma_{air} - 1} = \frac{1.4 \times 0.277}{1.4 - 1} \text{ J/g. K} = 0.9695 \text{ J/g. K}$$

$$(C_v)_{air} = \frac{R_{air}}{\gamma_{air} - 1} = \frac{0.277}{1.4 - 1} \text{ J/g. K} = 0.6925 \text{ J/g. K}$$

$$R_{Freon} = \frac{R_0}{\mathcal{M}_{Freon}} = \frac{8.314}{200} \text{ J/g. K} = 0.04157 \text{ J/g. K}$$

$$(C_p)_{Freon} = \frac{\gamma_{Freon}R_{Freon}}{\gamma_{Freon}-1} = \frac{1.1\times 0.04157}{1.1-1}\text{J/g. K} = 0.45727\,\text{J/g. K}$$

$$(C_v)_{Freon} = \frac{R_{Freon}}{\gamma_{Freon}-1} = \frac{0.04157}{1.1-1}\text{J/g. K} = 0.4157\,\text{J/g. K}$$

Now, $(C_p)_{mixture} = \dfrac{m_{air}(C_p)_{air} + m_{Freon}(C_p)_{Freon}}{m_{air} + m_{Freon}}$

$$= \frac{21\times 0.9695 + 60\times 0.45727}{21+60}\text{J/g. K}$$

$$= 0.59\,\text{J/g. K}$$

Now, $(C_v)_{mixture} = \dfrac{m_{air}(C_v)_{air} + m_{Freon}(C_v)_{Freon}}{m_{air} + m_{Freon}}$

$$= \frac{21\times 0.6925 + 60\times 0.4157}{21+60}\text{J/g. K}$$

$$= 0.4875\,\text{J/g. K}$$

Therefore, $\gamma_{mixture} = \dfrac{(C_p)_{mixture}}{(C_v)_{mixture}} = \dfrac{0.59}{0.4875} = 1.21$

Hence, the answer is (B).

Q16. Air-water vapour mixture having 100% relative humidity at 50°C is heated is obarically to 100°C in a closed system. If saturation pressure at 50°C is 12.352 kPa and at 100°C is 101.42 kPa, final relative humidity is

(A) 0% (B) 8.2% (C) 12.2% (D) 100%

Solution:

Relative humidity = vapour pressure at 100°C/ Saturation pressure at 100°C × 100

= 12.352/101.42 × 100%

= 12.2%

Hence, the answer is (C).

Common Data for Questions 17 and 18

Saturated vapour enters a turbine at a pressure of 2 bar and leaves the turbine at a pressure of 0.1 bar and a quality of 0.9. After condensation, saturated water al 0.1 bar is pumped into the boiler where it receives heat at a constant pressure of 2 bar. The pumping process can be considered to be isentropic. Use the data given in the following table to answer **Q.17** and **Q.18**.

Pressure (bar)	Temperature (°C)	Specific volume (m^3/kg)		Specific enthalpy (kJ/kg)		Specific entropy (kJ/kg-K)	
		v_f	v_g	h_f	h_g	s_f	s_g
2	120.23	0.001061	0.8857	504.68	2706.6	1.530	7.1271
0.1	45.81	0.001010	14.674	191.81	2584.6	0.6492	8.1501

Q17. Work done by the turbine is

(A) 276.5 kJ/kg
(B) 303.9 kJ/kg
(C) 335.8 kJ/kg
(D) 361.3 kJ/kg

Solution:

$h_2 = h_f + 0.9(h_g - h_f)$ = {191.81 + 0.9 × (2584.6-191.81)} kJ/kg = 2345.321 kJ/kg

$W = h_1 - h_2$ = (2706.6 - 2345.321) kJ/kg =361.3 kJ/kg

Hence, the answer is (D).

Q18. Heat addition in the boiler is

(A) 2000.9 kJ/kg
(B) 2514.6 kJ/kg
(C) 3028.2 kJ/kg
(D) 3554.5 kJ/kg

Solution:

Work supplied to the pump,

$W_p = v_f|_{0.1\ bar}(p_{boiler} - p_{condenser})$ = 0.001010 X (200-10) kJ/kg = 0.1919 kJ/kg

Hence, the enthalpy of steam at the inlet to the boiler

$h_4 = h_f|_{0.1\ bar} + W_p$ = (191.81+0.1919) kJ/kg = 192.0019 kJ/kg

Therefore, the heat transferred in the boiler

$Q = h_1 - h_4$ = (2706.6-192.0019) kJ/kg = 2514.6 kJ/kg

Hence, the answer is (B).

Common Data for Questions 19 and 20

An insulated piston-cylinder assembly having a paddle wheel, as shown in the adjacent figure, contains air (R = 287 J/kg-K, and C_v = 718 J/kg-K) of mass 4 kg. Both piston and paddle wheel can be considered as insulated and massless. Temperature and pressure of air inside the cylinder are 300 K and 100 kPa respectively. Ambient pressure is 100 kPa.

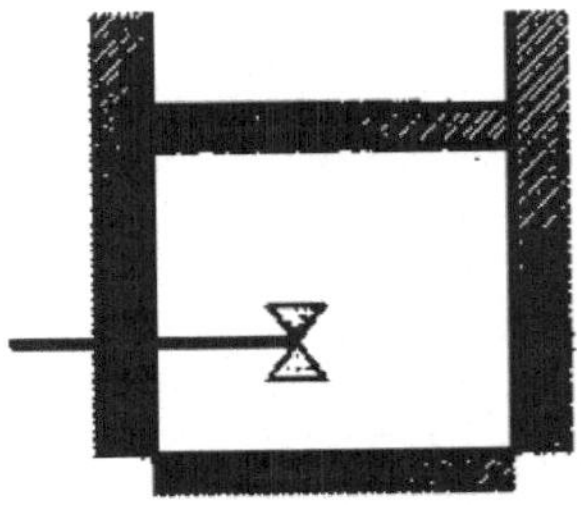

Q19. If the piston is locked in the fixed position and the paddle wheel delivers 75 kJ of work, final air temperature is

(A) 300 K (B) 318.7 K
(C) 320.6 K (D) 326.1 K

Solution:

The first law of thermodynamics for a closed system is

$Q = \Delta U + W.$

Since the present system is insulated, therefore , and hence

$\Delta U + W = 0$

$\Delta U = -W$

$4 \times 0.718 \times (T_2 - 300) = (-75)$

$T_2 = 326.1$ K

Hence, the answer is (D).

Q20. If the piston is free to slide without any friction when the paddle wheel delivers 75 kJ of work, final temperature of air in the cylinder is

(A) 305.2 K
(B) 309.3 K
(C) 312.6 K
(D) 318.7 K

Solution:

In this case the first law of thermodynamics will be

$\Delta U + p\Delta V + W_{paddle} = 0$

$\Delta U + \Delta(pV) = -W_{paddle}$

$\Delta(U + pV) = -W_{paddle}$

$\Delta H = -W_{paddle}$

$4 \times 1.005 \times (T_2 - 300) = (-75)$

$T_2 = 318.7$ K

Hence, the answer is (D).

Statement for Linked Answer Questions 21 and 22:

In a process industry, two different streams of water (to be considered incompressible) are available at 10°C and 90°C as shown in the figure. Mass flow rates of both the streams are 1 kg/s. Rather than wasting these resources, it is desired to connect a reversible Carnot engine that will continuously extract heat from the hot stream and supply part of it to the cold stream such that the exit temperature of both the streams T_f is identical. Heat capacity of water is 4.18 kJ/kg – K.

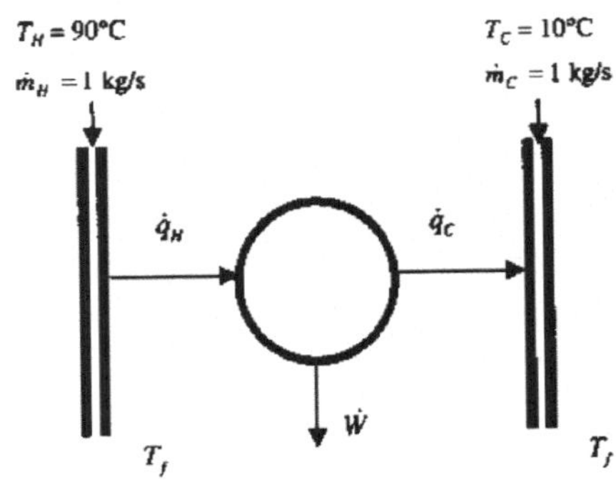

Q21. Value of T_f is

(A) 30°C (B) 42.5°C (C) 47.5°C (D) 50°C

Solution:

$$\frac{\dot{q}_H}{\dot{q}_C} = \frac{\frac{90+T_f}{2}+273}{\frac{10+T_f}{2}+273}$$

$$\frac{\dot{m}_H C_P (90-T_f)}{\dot{m}_c C_p (T_f-10)} = \frac{90+T_f+546}{10+T_f+546}$$

$$\frac{90-T_f}{T_f-10} = \frac{T_f+636}{T_f+556}$$

$$2T_f^2 + 636T_f - 10T_f - 6360 = 90T_f + 50040 - T_f^2 - 556T_f$$
$$2T_f^2 + 546T_f - 28200 = 0$$
$$T_f = 47.5°C$$

Hence, the answer is (C).

Q22. Work output W is

(A) 20.8 kW (B) 42.5 kW (C) 63 kW (D) 167 kW

Solution:

$$\dot{q}_H = \dot{m}_H C_p (90-T_f) = 1\times 4.18\times(90-47.5)\text{kW} = 177.65 \text{ kW}$$

$$\dot{q}_C = \dot{m}_C C_p (T_f-10) = 1\times 4.18\times(47.5-10)\text{kW} = 156.75 \text{ kW}$$

Therefore, the work done

$$\dot{w} = \dot{q}_H - \dot{q}_C = (177.65 - 156.75)\text{kW} = 20.9 \text{ kW}$$

Hence, the answer is (A).

GATE 2009
THERMODYNAMICS

Q. 1-8 carry one mark each

Q1. The ideal gas law is valid for:

(A) inert gases

(B) gases at high pressure and high temperature

(C) gases at low pressure and low temperature

(D) gases at low pressure and high temperature

Solution:

The answer is (D).

Q2. During the adiabatic saturation process

(A) the relative humidity increases but the specific humidity remains constant

(B) both the relative and specific humidity remains constant

(C) both the relative and specific humidity increases

(D) relative humidity decreases but the specific humidity increases

Solution:

The answer is (C).

Q3. For an ideal gas undergoing a throttling process 1-2, which of the following relationship holds?

(A) $T_1 = T_2$

(B) $\frac{P_1}{P_2} = \frac{T_1}{T_2}$

(C) $\frac{P_1}{P_2} = \left(\frac{T_1}{T_2}\right)^{\frac{\gamma}{\gamma-1}}$

(D) $\frac{P_1}{P_2} = \frac{T_2}{T_1}$

Solution:

In throttling process, enthalpy = constant

$\Delta h = 0$ (i)

For an ideal gas eq. (i) reduces to

$C_p \times (T_2 - T_1) = 0$

$T_1 = T_2$

Hence, the answer is (A).

Q4. A Carnot refrigerator operating between -1°C and 33°C has a cooling capacity of 1.6 kW. The power consumed by the refrigerator is

(A) 160 W (B) 178 W (C) 200 W (D) 1.8 kW

Solution:

$$COP = \frac{T_0}{T_k - T_0} = \frac{(-1+273)}{33-(-1)} = 8$$

Therefore, the power consumed = cooling capacity/COP= 1.6/8 kW = 0.2 kW = 200 W

Hence, the answer is (C).

Q5. An ideal gas undergoes expansion according to process $PV^{0.5}$ = constant. The temperature of the gas during the expansion process

(A) does not change

(B) increases

(C) decreases

(D) change depending on the initial condition

Solution:

For a polytropic process of an ideal gas obeying the relation we have

$$\frac{T_2}{T_1} = \left(\frac{P_2}{P_1}\right)^{\frac{n-1}{n}} \quad \ldots\text{(i)}$$

For the present case, n = 0.5.

Therefore, we have from eq. (i)

$$\frac{T_2}{T_1} = \left(\frac{P_2}{P_1}\right)^{-1} = \frac{P_1}{P_2} \quad \ldots\text{(ii)}$$

Now, for an expansion process $P_1 > P_2$. Therefore, from eq. (ii) $T_2 > T_1$.

Hence, the answer is (B).

Q6. Air (γ = 1.4) is compressed ideally from an initial state of 1 bar, 300 K to a final temperature of 600 K. The value of the pressure in bar is

(A) 2 (B) 3.7 (C) 7.2 (D) 11.3

Solution:

$(P_2/P_1) = (T_2/T_1)^{\gamma/(\gamma-1)}$

$P_2 = 1 \times (600/300)^{1.4/(1.4-1)}$ bar

P_2= 11.3 bar

Hence, the answer is (D).

Q7. On a *T-s* diagram, the slope of the constant volume line for an ideal gas is

(A)less than that of constant pressure line

(B) more than that of constant pressure line
(C) less than that of constant enthalpy line
(D) equal to that of constant enthalpy line

Solution:

For an ideal gas,

$$\left(\frac{\partial T}{\partial s}\right)_p = \frac{T}{C_p} \quad \text{(i)}$$

And,

$$\left(\frac{\partial T}{\partial s}\right)_v = \frac{T}{C_v} \quad \text{(ii)}$$

Since, $C_p > C_v$

hence from eqs. (i) and (ii)

$$\left(\frac{\partial T}{\partial s}\right)_v > \left(\frac{\partial T}{\partial s}\right)_p \quad \text{(iii)}$$

Hence, the answer is (B).

Q8. The thermal efficiency of an ideal Rankine cycle is less than that of Carnot cycle operating between the same maximum and minimum temperature limits, because

(A) heat addition does not take place at constant temperature
(B) the expansion process is not reversible and adiabatic
(C) heat rejection does not take place at constant temperature
(D) the compression process is not reversible and adiabatic

Solution:

The thermal efficiency of a power cycle increases if the temperature of heat addition is high. However, the heat addition in a Rankine cycle does not take place at a constant temperature. Therefore, the average temperature of heat addition in a Rankine cycle is less than the maximum temperature. Consequently, the efficiency of a Rankine cycle operating between given limits of maximum and minimum temperatures is less than that of a Carnot cycle.

Hence, the answer is (A).

Q.9-Q.24 carry two marks each.

Q9. Atmospheric air (R = 287 J/kg; γ = 1.4) at 1 bar and 25°C is compressed adiabatically to 2 bar and 105°C. Which of the following statement is correct?

(A) The process is possible but irreversible
(B) This process is possible and reversible
(C) The process is impossible
(D) The process is possible and isentropic

Solution:

For an adiabatic process, it is known from the second law of thermodynamics that

$$\Delta S \geq 0, \quad \text{(i)}$$

where the equality sign holds good for a reversible process and the inequality sign holds good for a reversible process.

Now, the mass of air be Therefore,

$$\Delta S = m\left\{C_p \ln\left(\frac{T_2}{T_1}\right) - R\ln\left(\frac{P_2}{P_1}\right)\right\}$$

$$= m\left\{\frac{\gamma R}{\gamma - 1}\ln\left(\frac{T_2}{T_1}\right) - R\ln\left(\frac{P_2}{P_1}\right)\right\}$$

$$= m\left\{\frac{1.4 \times 287}{1.4 - 1}\ln\left(\frac{105 + 273}{25 + 273}\right) - 287\ln\left(\frac{2}{1}\right)\right\} \text{ J/K}$$

$$= 39.93m \text{ J/K}$$

Since $\Delta S > 0$, therefore process is possible but irreversible

Hence, the answer is (A).

Q10. A pressure cooker contains saturated water-vapour mixture at 100°C with volume of vapour being eight times that of the volume of liquid. The specific volume of liquid and saturated vapour at 100°C are v_f = 0.001044 m³/kg and v_g = 1.6729 m³/kg, respectively. The quality of the mixture is

(A) 0.005 (B) 0.125 (C) 0.889 (D) 0.995

Solution:

Let the volume of liquid in the mixture be m³ and the volume of vapour V_g be m³. Therefore,

$$V_g = 8V_f \quad \text{(i)}$$

Therefore, the mass of liquid in the mixture, $m_f = \frac{V_f}{v_f} = \frac{V_f}{0.001044}$ kg (ii)

And, the mass of vapour in the mixture, $m_g = \frac{V_g}{v_g} = \frac{8V_f}{1.6729}$ kg (iii)

Therefore, the dryness fraction is given by $x = \frac{m_g}{m_g + m_f}$

$$= \frac{\frac{8V_f}{1.6729}}{\frac{8V_f}{1.6729} + \frac{V_f}{0.001044}}$$

$$= 0.005$$

Hence, the answer is (A).

Q11. An ideal gas (γ=1.39) flows in a pipeline at 450°C and 20 bar. A rigid, insulated and initially evacuated vessel is connected to the pipeline through a valve. The valve is now opened and the gas is allowed to fill the empty vessel. The final temperature of the gas in the vessel is

(A) 247°C (B) 450°C

(C) 625°C (D) 732°C

Solution:

$T_2 = \gamma \times T_1 = 1.39 \times (450 + 273) = 1004.97$ K = 732°C

Hence, the answer is (D).

Q12. An equi-molar mixture of nitrogen ($\gamma = 1.4$) and helium ($\gamma = 1.67$) is initially at 5 bar and 300°C. The mixture is expanded adiabatically to a pressure of 2 bar. The final temperature of the mixture is

(A) 149°C (B) 200°C (C) 250°C (D) 524°C

Solution:

$$(C_p)_{mix} = \frac{n_1(C_p)_1 + n_1(C_p)_2}{n_1 + n_2} = \frac{1}{2}\left(\frac{\gamma_1 R_0}{\gamma_1 - 1} + \frac{\gamma_2 R_0}{\gamma_2 - 1}\right) = \frac{R_0}{2}\left(\frac{1.4}{1.4-1} + \frac{1.67}{1.67-1}\right) = 2.996R_0$$

$$(C_v)_{mix} = \frac{n_1(C_p)_1 + n_1(C_p)_2}{n_1 + n_2} = \frac{1}{2}\left(\frac{R_0}{\gamma_1 - 1} + \frac{R_0}{\gamma_2 - 1}\right) = \frac{R_0}{2}\left(\frac{1}{1.4-1} + \frac{1}{1.67-1}\right) = 1.996R_0$$

Therefore, $\gamma_{mix} = 1.5$

$$\text{Therefore, } T_2 = T_1\left(\frac{P_2}{P_1}\right)^{\frac{\gamma_{mix}-1}{\gamma_{mix}}} = (300 + 273) \times \left(\frac{2}{5}\right)^{\frac{1.5-1}{1.5}} \text{K} = 422 \text{ K} = 149°\text{C}$$

Hence, the answer is (A).

Q13. A heat engine E_1 operates between an infinite reservoir at 800°C and a body B. The temperature of the body B remains constant at 550°C. Heat transferred to the engine E_1 is 900 kJ and the work output is 200 kJ. Another engine E_2 operates between the body B and the atmosphere at 27°C. Heat rejected to the atmosphere is 350 kJ. The thermal efficiency of the engine E_2 is

(A) 0.39 (B) 0.5 (C) 0.61 (D) 0.635

Solution:

Q_1 = 900 kJ, W = 200 kJ, Q_3 = 350 kJ

$Q_1 = W + Q_2$ => Q_2 = (900-200) kJ = 700 kJ

Thermal efficiency of $E_2 = 1-(Q_3/Q_2) = 1-(350/700) = 0.5$

Hence, the answer is (B).

Q14. A gas turbine power plant operates with air (γ=1.4) between 1 bar and 20 bar. The maximum thermal efficiency (in %) for the corresponding air standard cycle is

(A) 30 (B) 36.7 (C) 48.2 (D) 57.5

Solution:

$$\eta_{Braytom} = 1 - \frac{1}{r_p^{\frac{\gamma-1}{\gamma}}} = 1 - \frac{1}{20^{\frac{1.4-1}{1.4}}} = 0.575$$

Hence, the answer is (D).

Q15. The saturation pressure of water at 100°C and 105°C are respectively 101.3 kPa and 120.8 kPa. Taking the molecular weight of water as 18, the latent heat of water in kJ/kg at 102.5°C is approximately equal to

(A) 2290 (B) 1250 (C) 820 (D) 330

Solution:

$$\text{In}\left(\frac{P_2}{P_1}\right) = \Delta H \times \frac{M_{H2O}}{R0} \times \left(\frac{1}{T_2} - \frac{1}{T_1}\right)$$

$$\text{In}\left(\frac{120.8}{101.3}\right) = \Delta H \times \frac{18}{8.314} \times \left(\frac{1}{378} - \frac{1}{373}\right)$$

ΔH = 2290 kJ/kg

Hence, the answer is (A).

Q16. An engine reversibly receives 1200 J of heat at 900 K. After rejecting heat to the ambient at 300 K, it develops 600 J of work. The irreversibility in joules is equal to

(A) 600 (B) 400 (C) 200 (D) Zero

Solution:

Heat rejected into the sink, $Q_2 = Q_1 - W$ = (1200 – 600) J = 600

Irreversibility $= T_0 \Delta S$ = 300 × (600/300 – 1200/900) J = 200 J

Hence, the answer is (C).

Q17. Saturated liquid water at 0.4 MPa and 1000 kg/hr of steam at 0.4 MPa and 300°C enter steadily into an insulated mixing chamber. At 0.4 MPa, the enthalpy of saturated liquid and saturated vapour are 604.73 and 2738.53 kJ/kg respectively; also, the enthalpy of the superheated steam at 300°C is 3066.75 kJ/kg. The quality of water-vapour mixture exiting in the chamber is 0.9. The mass flow rate of saturated liquid water in kg/hr is

(A) 182 (B) 282 (C) 382 (D) 1000

Solution:

$h_3 = h_f + x^*(h_g - h_f)$ = {604.73 + 0.9 × (2738.53 - 604.73)} kJ/kg = 2525.15 kJ/kg

Now, mass balance gives:

Further, the energy balance for the mixing process gives

$$\dot{m}_1 + \dot{m}_2 = \dot{m}_3$$

$$\dot{m}_1 h_2 + \dot{m}_2 h_2 = \dot{m}_3 h_3$$

$$\dot{m}_1 \times 604.73 - 1000 \times 3066.75 = (\dot{m}_1 + 1000) \times 2525.15$$

$$\dot{m}_1 = 282 \text{ kg/hr}$$

Hence, the answer is (B).

Q18. A certain quantity of gas undergone the polytropic process, $PV^{1.3}$ = constant, from an ideal state of 1.5 MPa and 0.09 m³ to the final pressure of 7.5 MPa. The work done by the gas in kJ is

(A) -217 (B) -200 (C) 200 (D) 217

Solution:

$P_1V_1^{1.3} = P_2V_2^{1.3}$

$V_2 = 0.09 \times (1.5/7.5)^{1/1.3} = 0.02609$ m³

$W = (P_1V_1 - P_2V_2)/(n - 1) = -202.25$ kJ

Hence, the answer is (B).

Common Data for Questions 19 and 20:

Saturated water vapour enters an adiabatic turbine at 0.8 MPa and leaves at 0.1 MPa. The mass flow rate of water vapour is 25 kg/s. Use the following data table to answer the questions 19 and 20.

Pressure (MPa)	Temperature (°C)	Specific enthalpy		Specific entropy	
		h_f (kJ/kg)	h_g (kJ/kg)	s_f (kJ/kg K)	s_g (kJ/kg K)
0.8	170.43	722.11	2769.10	2.0462	6.6628
0.1	99.63	417.46	2675.50	1.3026	7.3594

Q19. The quality of the steam at the exit of the turbine after an isentropic expansion is

(A) 0.47 (C) 0.72 (B) 0.88 (D) 0.94

Solution:

$s = s_f + x(s_g - s_f)$

$6.6628 = 1.3026 + x \times (7.3594 - 1.3026)$

$x = 0.88$

Hence, the answer is (C).

Q20. If the steam leaves the turbine as saturated vapour, the power produced by the turbine in kW is

(A) 1640 (B) 2030

(C) 2340 (D) 8830

Solution:

Power produced by turbine

$= \dot{m}(h_g |_{0.8\,\text{MPa}} - h_g |_{0.1\,\text{MPa}}) = 25 \times (2769.10 - 2675.50)\ \text{kW} = 2340\ \text{kW}$

Hence, the answer is (C).

Common Data for Questions 21 and 22

The flow rate of refrigerant R-12 in a refrigerator is 0.03 kg/s. The refrigerant enters the compressor as saturated vapour at 150.9 kPa. After adiabatic compression, it enters the condenser as superheated vapour at 500 kPa and 100°C and leaves the condenser as saturated liquid at the same pressure. Use the following data to answer the questions 21 and 22.

Pressure (kPa)	Temperature (°C)	Specific enthalpy	
		h_f (kJ/kg)	h_g (kJ/kg)
150.9	-20	17.82	178.74
500	15.6	50.64	195.01

For the superheated vapour at 500 kPa and 100°C, h = 252.05 kJ/kg.

Q21. The refrigeration effect in kW is

(A) 1.71 (B) 3.84 (C) 4.33 (D) 4.83

Solution:

Refrigeration effect = (178.74 – 50.64) × 0.03 kW = 3.84 kW

Hence, the answer is (B).

Q22. The actual power of input to the compressor in kW is

(A) 0.49 (B) 0.99 (C) 1.71 (D) 2.2

Solution:

Actual power i/p to the compressor = (252.05-178.74) × 0.03 kW = 2.2 kW

Hence, the answer is (D).

Statement for Linked Answer Questions 23-24:

An insulated vertical cylinder encloses 0.1 kg of argon (Ar) with the help of a frictionless non-conducting piston as shown in the figure. The mass of the piston is 5 kg and it initially rests on the bottom of the cylinder. The cylinder is connected to a nitrogen (N_2) tank at 100 bar through a pipeline fitted with a valve. The valve is opened and nitrogen is slowly admitted into the cylinder. During this operation, the piston is lifted through a height of 10 cm by the nitrogen gas. The initial pressure and temperature of argon gas are 100 kPa and 300 K respectively. The final temperature of argon is 320 K. For argon, $C_p = 520$ J/kg K and $C_v = 312$ J/kg K.

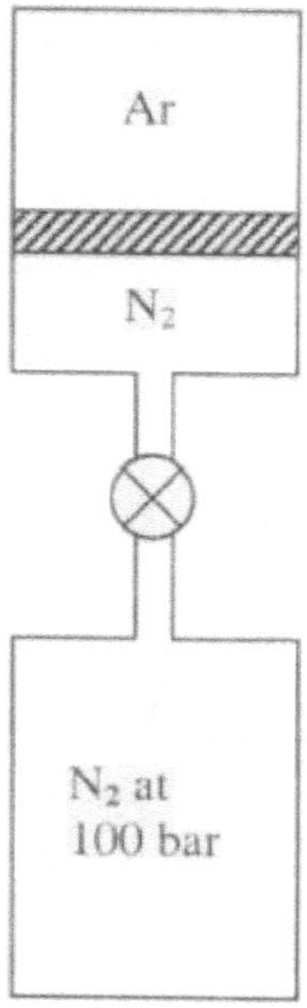

Q23. The work done by argon in kJ during the process is

(A) 10 (B) 1.041 (C) -0.624 (D) -1.041

Solution:

$W = -\Delta U = -mC_v(T_2 - T_1) = -0.1 \times 312 \times (320-300) = -0.624$ kJ

Hence, the answer is (C).

Q24. The work done by nitrogen in kJ during the process is

(A) 1.046 (B) 0.629 (C) -1.046 (D) -10

Solution:

Considering the acceleration due to gravity = 10 m/s², the work done by nitrogen = The Work done by nitrogen on argon + Work done by nitrogen in lifting the piston

$$= \left(0.624 + \frac{5\times10\times0.1}{1000}\right) \text{kJ}$$

$$= 0.629 \text{ kJ}$$

Hence, the answer is (B).

9 789359 197838

Printed at Repro India Ltd.